The Origin and Early Evolution of Life

The Origin and Early Evolution of Life

Tom Fenchel
Professor of Marine Biology
University of Copenhagen

OXFORD
UNIVERSITY PRESS

Great Clarendon Street, Oxford OX2 6DP

Oxford University Press is a department of the University of Oxford.
It furthers the University's objective of excellence in research, scholarship,
and education by publishing worldwide in

Oxford New York

Auckland Bangkok Buenos Aires Cape Town Chennai
Dar es Salaam Delhi Hong Kong Istanbul Karachi Kolkata
Kuala Lumpur Madrid Melbourne Mexico City Mumbai Nairobi
São Paulo Shanghai Taipei Tokyo Toronto

Oxford is a registered trade mark of Oxford University Press
in the UK and in certain other countries

Published in the United States
by Oxford University Press Inc., New York

First published 2002

A catalogue record for this title is available from the British Library.

Library of Congress Cataloging in Publication Data

Fenchel, Tom.
Origin and early evolution of life/Tom Fenchel.
Includes bibliographical references (p.).
1. Life–Origin 2. Evolution (Biology) I. Title.

QH325 .F42 2002 576.8'3–dc21 2002070193

ISBN 0 19 852533 8 (alk. paper: pbk.)
ISBN 0 19 852635 0 (alk. paper: hbk.)

10 9 8 7 6 5 4 3 2 1

Typeset by Cepha Imaging Pvt Ltd, India
Printed on acid-free paper by
T. J. International Ltd, Padstow

Preface

I have lectured on the early evolution of life for several years as part of an evolution course for undergraduate biology students at the University of Copenhagen; these lectures constitute the backbone of this book. It was originally published in Danish; while translating it into English I have made a few changes and additions.

I cannot believe that anyone would be so naïve to think that a title containing the phrase '*origin of life*' means that a book actually explains how life originated—because this is not known. But preoccupation with the problem gives insight into what life is. When it comes to the evolution of life and the biosphere after the first bacterial cell arose and until the origin of higher organisms—a period that covers most of geological time—a lot can actually be said, although there are still many unresolved problems. While several excellent books exist that describe particular aspects of the subject matter, none of them attempt to provide a complete picture of our current understanding of the early evolution of life, simultaneously drawing evidence from many disparate fields ranging from molecular genetics to geology; hence this book. It is addressed to biology students and colleagues, but it should be accessible to anyone with some background in science—and so I hope that the book will contribute to a more widespread interest in and understanding of biology.

The following colleagues have helped in locating and lending me photographs used in the book: Dr Stanley M. Awramic, University of California, Santa Barbara; Professor Donald E. Canfield, Institute of Biology, University of Odense; Professor Bo Barker Jørgensen, Max Planck Institute for Marine Microbiology, Bremen; Dr Michael Kühl, Marine Biological Laboratory, University of Copenhagen; and Dr Jacob Larsen, Department of Algae and Fungi, University of Copenhagen. The Danish version of the book was improved through suggestions and comments from Donald E. Canfield, Dr Ronnie Glud, Marine Biological Laboratory, University of Copenhagen, and Professor Kaj Sand-Jensen, Freshwater Biological Laboratory, University of Copenhagen. I am particularly grateful to Professor Freddy B. Christiansen, Bioinformatics Research Center, University of Aarhus; his comments and suggestions have substantially improved the book. Finally, I am very grateful to Professor Bland J. Finlay, CEH Windermere, UK for improving my English as well as for suggestions concerning scientific substance. Traditionally I should add that all errors and shortcomings remain my responsibility.

Helsingør
December 2001

Tom Fenchel

Contents

Chapter 1

Introduction

This book is about the development of life from its origin and until multicellular plants, fungi, and animals arose—corresponding approximately to the time period from 4 to 0.6 billion years ago. If the reader expects a complete explanation for the origin of life and its earliest differentiation then she or he will be disappointed. First of all these questions will never be fully answered because the origin of life is a historical event. Plausible ideas about mechanisms that possibly led to the origin of life may be suggested and perhaps, in some cases at least, tested experimentally. But it will never be possible to prove that this is how it happened. Also, there are huge gaps in our understanding of the earliest evolution of life.

Nevertheless it may still be worthwhile devoting time to the problem and to write books about it. First of all, it is the most important question for biology and perhaps for science in general. Also, and even though no complete answers can be given, preoccupation with the topic leads to a more profound understanding of what life is. Finally, something sensible can still be said about the topic, and much new insight has been gained during the last 20–30 years. This is mainly due to developments in molecular genetics, an increased understanding of microbial physiology, and progress in geology.

Some important results are:

1. Life originated very early during Earth's history (perhaps about four billion years ago) and during the first 3.4 billion years it was represented only by microbes. About 50 years ago only a few, uncertain traces of life older than about 600 million years were known.
2. All life is related in a genealogical sense—we know only one kind of life. In spite of the apparent diversity of life, the fundamental mechanisms and processes characterizing life are quite uniform and must have been established very early. There are three main groups (domains) of life: eubacteria, archaebacteria, and eukaryotes. The two former groups are similar with respect to their organizational level and are collectively called prokaryotes or just 'bacteria'. Eukaryotes seem to represent a leap in complexity relative to bacteria. All fundamental biochemistry is common to the three groups, however, and must have been established before they diverged.
3. During the last 3.5–4 billion years life has been a driving force in biogeochemical cycling on the surface of Earth and it has profoundly influenced the chemistry of the atmosphere and the oceans.

The topic of the book invites speculation. Speculation—or better, perhaps, 'knowledge-based intuition'—is not strange to science. Rather it is a necessary tool for choosing the right experiments and for collecting meaningful data in nature. In order to prevent speculation degenerating into pseudo-science or science fiction, however, speculations must be constrained. In our context this means that speculations must be in agreement with laws of physics and chemistry, with the principles of evolution through natural (Darwinian) selection, with our knowledge on the structure and function of extant organisms, and with geological data. The subject matter of this book is very broad, covering large parts of biology and venturing into biochemistry, molecular genetics, geology, and geochemistry—together all this would cover many volumes each with the thickness of a telephone directory. It will therefore be necessary to simplify the treatment of some aspects. To do so without introducing errors or misleading statements requires insight and expertise. In my trade one is usually assumed to be a specialist in a relatively narrow field. And so, by taking this broad approach to the topic, I run the risk of disclosing a lack of authoritative grip on some topics.

Certain themes are recurrent in the book. These are:

1. *The 'What came first—the chicken or egg?' problem.* More concretely: The genome of organisms contains the necessary information for producing complex protein molecules (enzymes). These, in turn, are necessary for the translation of genetic information and for replication of the genome. How could such a system in which the two components are mutually dependent ever arise?
2. *The unit of natural selection.* Darwin assumed—in most contexts quite correctly—that the driving force of evolution is selection of individual organisms on the basis of their fitness. Fitness may be considered a measure of the individual's adaptation to its environment and it is basically a measure of the number of offspring that survives to reproductive age. Fitness has many components such as competitive ability relative to its conspecifics, avoidance of being eaten by predators, ability to survive detrimental environments, etc. In some contexts it is, however, more natural to speak about the fitness of individual genes. But there are also examples of 'altruism' in the sense that some individuals sacrifice themselves for other, usually closely related, individuals. In such cases the unit of selection seems to be a group of individuals. Our ideas about evolution of life imply the integration of self-replicating molecules into a genome, horizontal transfer of genes from one individual to another, that the origin of eukaryotic cells implied the integration of endosymbiotic bacteria, and that multicellular organisms arose from cell colonies. At every level it is necessary to explain how such consortia could become stable units for natural selection.
3. *The evolutionary process depends on the available tools.* Nature is imperfect and biological evolution is an opportunistic process. Evolutionary change is contingent on the genetic variation at hand, and this variation is a result of random mutations without functional direction. But evolutionary change is also constrained by already established structures and functions. The fact that our oesophagus and pharynx cross each other is an example. It is an unfortunate construction of which we are reminded every time something gets stuck in our throat. The reason for this anatomical detail is that we are descended from

fish that once lived in oxygen-poor water. They developed the habit of swallowing air at the surface and then extracted oxygen through the intestinal wall (this adaptation is still found in many tropical freshwater fish). The system was later improved through special extensions from the intestine in the form of lungs (cf. modern lungfish). This accessory system for respiration was the precondition for the later evolution of terrestrial vertebrates. No creator, however, could have though of such an idiotic 'plumbing' for land-living vertebrates. Structures and functions of extant organisms reflect past evolutionary origins because evolution is based on the principle of available tools. This must also apply at the most fundamental level. The genetic code, fundamental biochemical pathways, etc., must also be based on earlier structures and functions and somehow represent keys to the understanding of the earliest start of life. It is only that in many cases we have not yet been able to interpret them.

It would be useful to mention a few aspects of evolutionary biology that sometimes lead to misunderstandings. Among these is the use of metaphors. Evolutionary biology has always had problems with metaphors because it seems difficult to avoid teleological implications. Adaptations are often explained in terms such as 'elephants have a trunk so that they can pull up trees'. This is, of course, an inappropriate way to explain why ancestors of elephants over time developed a trunk from a typical mammal nose, but a stringent explanation in terms of evolutionary theory would be tedious and sermonizing. Molecular genetics has also contributed to a number of metaphors such as the genetic 'code' and genetic 'information', that strictly speaking are misleading. The term 'evolution' itself is also somewhat ambiguous. Lots of things 'evolve': personalities and dramas, and in nature volcanoes, stars, and chemical processes also evolve. Such phenomena always have some more or less complicated causal reasons and the term 'evolution' is unavoidable in many contexts. In the context of Darwinian evolution the term is much more delimited, meaning only that evolutionary lineages change over time due to mutations, horizontal gene transfer, natural selection, and genetic drift. Metaphors seem to be unavoidable, but it is necessary to recognize their limitations.

The book discusses the evolution from simple self-replicating molecules via the simplest known organisms to large and complex organisms. In this way there is a risk of supporting a common, but false notion that biological evolution necessarily leads to increased complexity—rather like the old evolutionary trees that displayed an amoeba at the bottom and *Homo sapiens* at the top. Evolution, however, only favours Darwinian fitness and there are lots of examples of secondary simplification of organisms such as cavefish without eyes, snakes without limbs, etc. Evolution of parasites often leads to a drastic simplification of organisms in that the parasite can exploit a number of host functions and so dispose of its own similar resource-requiring functions. The ultimate evolution of a parasite is to become a virus that basically consists only of genetic material (RNA or DNA) and a protein coat that is necessary for entering host cells. Evolution does not have complexity—or anything else—as an end-goal.

It is not meaningful to apply the term 'successful' to any taxonomic group of organisms. It is often stated that, for example, insects represent an especially successful group of animals because they include so many species, or because it is hard to avoid stepping on them in all kinds of places. But the only criterion for success that possibly could give any meaning would be that a given lineage has not (yet) gone extinct. If a certain group should be considered successful in terms of their significance for the chemical composition and transformations of the biosphere or for their total biomass on Earth, then the prize should go the bacteria. These organisms have lived on Earth twice as long as any other kind of life, and most basic bacterial phenotypes have probably been around through most of Earth's history.

The structure of the book. The 14 chapters of the book are organized as follows. Chapter 2 presents only a short overview of geological time in order to provide a feeling for the time frame of the events discussed later in the book. It appears to be a dilemma that 'modern' bacteria were present very early in Earth's history leaving a relatively short time span from the origin of life to the arrival of a cell. Later major transitions and adaptive radiations (eukaryotic cells and much later multicellular animals and plants) also took place over what are in a geological time-scale short periods of time. Chapter 3 provides a brief account of the history of the topic running up to recent ideas and experimental results that still play a role in discussions of the origin of life. Chapter 4 discusses the central problem of what characterizes life and how it can be defined. Chapters 5 and 6 deal with the current status of our understanding of the origin of life. Most emphasis is on experimental results with self-replicating RNA molecules and on ideas of how such systems could possibly evolve into a real cell. The conclusion is that by and large we are left with many more questions than answers.

All organisms depend on energy and on building blocks so that they can grow and multiply. Chapter 7 is devoted to the origin and evolution of biochemical pathways and cycles, especially based on current understanding of bacterial energy metabolism. Chapter 8 deals with the origin of the eukaryotic cell; there is an apparent discontinuous leap in complexity as compared with bacteria. In part this can be explained in terms of eukaryotic cells as symbiotic consortia of bacteria, but several problems remain. The following chapter briefly discusses the origin of multicellularity. This chapter is in no way comprehensive, but it serves to delimit the main theme of the book. Summarizing so far, evidence shows that all fundamental cellular functions were developed very early and they are basically still common to all extant forms of life. Further evolution of complexity has mainly been based on a modular principle.

Chapter 10 will first briefly discuss sexual processes and their evolutionary significance among eukaryotes. In this context the species concept for microbes will be discussed. The usual presentation of evolutionary processes and of speciation rests mainly on out-breeding sexual animals and plants. The evolution of prokaryotes (and primitive eukaryotes) is evolution without sex and the implications of this are explored.

Life originated in an anoxic world. Although oxygen-producing cyanobacteria probably arose very early during evolution, the atmosphere and especially the seas maintained a very low oxygen tension throughout a large part of the Precambrian. All existing life still shows signs of anaerobic origin. These aspects are treated in Chapter 11.

Molecular genetics has provided tools that, in principle at least, can unravel the genealogical relatedness for all kinds of organisms, and construct an evolutionary tree that comprises all groups of organisms. These molecular trees have provided a totally new view of the evolution of life—they have also raised a number of new questions as discussed in Chapter 12.

Chapter 13 is devoted to the geological evidence for the earliest evolution of life. Palaeontology has during the last 40–50 years revealed Precambrian fossils and other traces of early life. Together with extant analogies, this has allowed for some understanding of early microbial communities. Other geological evidence provides some insight into the development of biogeochemical cycling and the chemical development of the atmosphere and the seas. The last chapter returns to the question of why only bacteria were represented through half of the entire period there has been life on Earth, and why eukaryotes and multicellular organisms occurred so late. It is suggested that these events required particular levels of atmospheric oxygen, and so the evolution of higher life forms was contingent on the biogeochemical peculiarities that allowed for accumulation of atmospheric O_2.

The book is not referenced, but at the end, some suggestions for further reading are provided. These references should also make it possible for readers to trace the original literature on which the book is based. Finally, there is a glossary that hopefully will make it easier to read the book.

Chapter 2

The geological time frame

From the beginning of the 19th century it became possible to establish a relative geological time-scale on the basis of the layering of different types of rock. A given stratum could be expected to be younger than underlying strata, and older than overlying ones. Some rocks may be sedimentary—deposited in water to become sandstone, shale, or limestone. Metamorphosed rocks resulted from the melting and subsequent solidification of other types of rocks. Igneous rocks were solidified lava or magma.

The naming of the geological periods also largely derives from the 19th century. Fossils in sedimentary rocks could characterize different periods. It became clear that the geological strata represented a time period that far exceeded Biblical tradition, according to which the world is about 6000 years old. But an absolute time-scale was not available. Different methods such as comparing the thickness of layers of sedimentary rocks with known sedimentation rates in the sea, or comparing the salt content of the sea with transport rates of dissolved salts into the sea from rivers yielded variable and unreliable estimates of geological age. Lord Kelvin tried to estimate the ages of the Earth and sun on the basis of their heat production and with the assumption that it was caused by the gravitational contraction during their formation. This provided improbably low estimates. He could not, of course, know that sun's heat generation is due to nuclear fusion processes or that Earth's heat generation is caused by radioactive decay. But for some time Lord Kelvin did cause problems for evolutionary biology and for geology.

Using decay constants for naturally occurring radioactive isotopes, it is today possible to make relatively precise absolute age determinations. In principle the approach is similar to the C-14 method used to age archaeological objects, in that the rate constant of radioactive decay is independent of external conditions such as temperature. Useful isotopes for geological age determination must, however, have much longer half-lives than that of C-14. Among such isotopes, uranium-238 has a half-life of 4.51×10^9 years; and via some intermediate products, it decays to Pb-206. Other useful isotopes are U-235 and Th-232 that decay to other isotopes of lead. In principle, the age of a rock can then be determined through the ratio of Pb-206 and U-238 as measured by mass spectrometry. In practise it is necessary to compensate for Pb-206 that might have been present in the lava before it solidified. Not all types of rocks can be aged by this method; it is in particular igneous rocks that can be used. The age of, for example, a layer of limestone can then be approximated by the ages of over- and underlying layers of lava.

Figure 2.1 shows the time-scale for the development of Earth with an indication of the first appearance of some of the main groups of organisms. More details on the

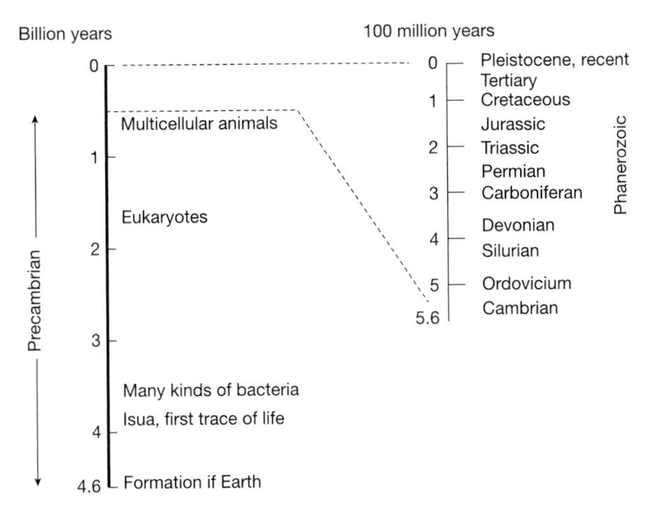

Fig. 2.1 The geological time frame. More details regarding the Precambrian are shown in Figure 13.2.

geological evidence for the evolution of life are provided in Chapter 13. Here we are concerned only with the time periods that seem to separate major evolutionary events.

The oldest known crystalline rocks are about four billion years old. The oldest sedimentary rocks are from the west coast of Greenland (Isua) and are about 3.8 billion years old. The reason why no older rocks are known is that Earth is geologically active. This is reflected in ocean floor spreading and its subsequent subduction, in volcanism, mountain building, and subsequent erosion of exposed rocks. Older geological deposits have therefore to a large extent been destroyed and the number of available deposits decreases strongly with their age. Also, during the first 200–500 million years of Earth's existence, it was exposed to impacts from meteorites and comets, as we can still see on the moon. The estimated age of the solar system (about 4.6×10^9 years) is therefore based on the ageing of meteorites and the oldest rocks found on the moon.

The last 544 million years of Earth's history is called the Phanerozoic and it is introduced by the Cambrian period. Until about 50 years ago the Phanerozoic corresponded to the period from which signs of life were known (the oldest remains of animals are about 590 million years old and multicellular algae extend another 10 million years back in time). Multicellular organisms have therefore inhabited Earth for less than one-seventh of geological time. From the Cambrian and onwards we know the evolutionary history of animals and plants in considerable detail. During the Cambrian, animals diverged into most of the invertebrate groups that we know today. The first vertebrates in the form of primitive fish appeared in the Silurian, a period in which land was also colonized by life. The first amphibians invaded land during the Devonian, and the first reptiles appear towards the end of Carboniferous together with the first seed plants. The Jurassic and the Cretaceous (190–65 million years ago) are the time of dinosaurs and the appearance of flowering plants and birds. In the beginning of the Tertiary, most modern mammal groups appear. The last two million years, up until about 10 000 years ago, is referred to as the Pleistocene and has been characterized by several ice-ages. Modern humans appeared about 50 000 years ago.

The period prior to the Cambrian (that includes six-sevenths of Earth's history, or about four billion years) is referred to as the Precambrian. Two subdivisions are recognized: the older Archaean and the more recent Proterozoic (the first 600 million years of Earth's existence is often referred to as the Hadean, and is the period from which we have no remains).

It must be assumed that life could not have developed prior to 4.4 billion years ago due to the intensity of meteorite impacts, and many believe that this applied up to about four billion years ago, because the oceans would regularly have been boiled away as a consequence of impacts of very large meteorites or asteroids. To what extent this applies is not really known. The 3.8 billion-year-old Isua deposits include, although somewhat metamorphosed, sedimentary rocks. This shows that the Earth already had continental masses, oceans, and a hydrological cycle. These rocks also contain carbon in the form of graphite grains. The ratio of stable carbon isotopes in these graphite grains suggests that they derive from organic matter that had been generated by some sort of photosynthetic process (see Chapter 13). This would indicate that life was already established at that time. It is quite certain that by some 300 million years later—that is 3.5 billion years ago—there existed a diverse bacterial biota possibly including cyanobacteria (previously referred to as blue-green 'algae'). The documentation for this is based on the finding of fossils from Western Australia and from South Africa. The first convincing traces of eukaryotic unicellular organisms are about 1.7 billion years old. Bacteria have thus been the only kind of life for about half of the time that there has been life on Earth. The first multicellular organisms appear, as already mentioned, about 600 million years ago.

All of this poses some fundamental questions. If life could not have arisen earlier than 4.4 or perhaps only four billion years ago, and since we have evidence of life that existed 3.8 billion years ago, then there is only a window of 600 or perhaps only 200 million years from the origin of life to the evolution of the first bacterial cells. If oxygen-producing photosynthetic bacteria appeared at least 3.5 billion years ago and this suggests that all extant basic types of bacterial metabolism had already developed by then. In this connection it should of course be noted that 600, or even 200 million years are enormous spans of time: 600 million years, after all, corresponds to more than the Phanerozoic period. We may also assume that for the first cells, the world was 'open' in the sense that different potential types of metabolism and other adaptations could evolve relatively fast. That the sea could quickly be filled with life can be seen from the following hypothetical situation. If the first cell (with a volume of 1 μm^3) was capable of a division every 24 hours, and that there were large amounts of unused resources, then it would take only a little more than 100 days for the descendants to cover the Earth's surface with a 1 cm thick layer. It is not unlikely that the time between the first strides of molecular evolution and the emergence of the first real bacterial cell was relatively short, possibly only 1–10 million years.

It is remarkable that the other great transitions and subsequent adaptive radiations—the emergence of the eukaryotic cell and, a little less than one billion years later, the emergence of multicellular organisms—also happened during relatively short periods of time. These periods seem to have been separated by long periods of time with less radical evolutionary changes.

The apparently sudden appearance of animals in the geological record also caused Darwin some concern. In *The origin of species* (1859) he dealt with a number of difficulties for his theory of evolution. One section: *On the sudden appearance of groups of allied species in the lowest known fossiliferous strata*, is dedicated to the problem of the absence of Precambrian fossils. He found it disturbing that Cambrian deposits, apparently representing the oldest signs of life, nevertheless contained advanced invertebrate animals such as nautiloid cephalolopods, trilobites, etc. Darwin's contemporaries knew of no signs of Precambrian life (except some indirect chemical evidence and the Canadian finding of '*Eozoon*' that has subsequently been shown to be structures of mineral origin). Darwin concluded correctly that the Cambrian fauna represented species that were too advanced and specialized to be ancestors of all later arthropods and cephalopods and that some earlier evolution must have taken place.

Later it was argued that animals first developed carbonate skeletons and shells in the beginning of the Cambrian. Such structures are usually the pre-condition for recognizable fossils; soft parts are preserved only under very special conditions. In the middle of the last century traces of a rather rich fauna of soft-bodied animals have been found including jellyfish-like coelenterates, annelids, and some forms that cannot be assigned to any particular animal group. This fauna lived during the period from 590 to 550 million years ago and is called the Ediacara fauna. It was first recorded from Southern Australia, but has now also been found elsewhere. It is probable that invertebrates existed even earlier, but that these were small, soft-bodied forms. Today there are many tiny invertebrates ('meiofauna') such as nematodes, rotifers, and turbellarians, some of which measure less than 100 μm. Such forms are unlikely to leave fossils. Molecular evolutionary trees for the animal kingdom also provide evidence that the main groups of invertebrates diverged even earlier than the Ediacara fauna.

Nevertheless, 'Darwin's dilemma' remains a real problem and not an artefact resulting from failure to find fossil evidence. If one had looked for the right traces, even in Darwin's time, it would have been possible to find much evidence of Precambrian life—but this would all have consisted of microbes. Today we know that microbial life thrived at least three billion years before the dawn of the Cambrian. So why do larger organisms appear so late, and then, over a short period of time, diverge into all the main groups of extant animals?

Chapter 3

Early ideas on the origin of life

Spontaneous generation

According to the Judeo–Christian tradition, the world and all living things originated in a singular event of creation. Today, all enlightened people in the western world consider Genesis as a myth without physical reality (so-called creationists in USA are excluded from the category of enlightened persons). But in earlier centuries the Church insisted on a literal interpretation of the biblical tradition, and its resistance to the theory of evolution continuing up until recent times exemplifies this.

With this background it seems paradoxical that the reality of *spontaneous generation* was a popular belief through the centuries and that it remained a respectable scientific hypothesis until around 1870. Spontaneous generation means that life originates spontaneously from non-living matter. Popular examples included the beliefs that eels form in the muddy lake bottom, that mice come from old rags in the attic, and maggots from decaying meat. In a sense these beliefs were simply empirical observations: where eels came from remained a mystery until about 70 years ago, and that mice are formed from old rags and maggots from decaying meat is something anyone could see with their own eyes. The idea of spontaneous generation can be traced back to Greek philosophers; Aristotle, for example, believed that life was generated in slime, compost, and mud under the influence of moisture and heat.

The Italian physician Fransesco Redi showed in 1668 that maggots come from eggs that blowflies deposit on meat. When he wrapped meat in gauze, flies could no longer deposit their eggs and no maggots developed. During this period, increasing knowledge of the life cycles of animals meant that ideas about spontaneous generation of macroscopic organisms were abandoned in learned circles. Towards the end of the 17th century, however, the Dutch amateur microscopist Antonie van Leeuwenhoek discovered microbes including bacteria and many kinds of protozoa. Over a period of 50 years he sent his observations to the Royal Society in London where they were published. Leeuwenhoek himself tended to believe that his microbes always derived from 'germs' and thus did not appear spontaneously. But this did not convince many other people. If dried plant remains were boiled in water and allowed to cool (an 'infusion', hence the term 'infusoria' for ciliated protozoa) it did not take long before microbes were swarming in the liquid. The Italian abbot and first experimental biologist Lazzaro Spallanzani carried out experiments in the middle of the 18th century to clarify the problem. He boiled infusions and immediately sealed the containers. These did not develop life and he concluded that microbes do not appear spontaneously, but that they must be able to spread through air. Many contemporaries did not accept the results of Spallanzani and of other 'anti-spontanists'.

Among the sceptics the French biologist Buffon believed that organic matter contained a 'vital principle' and even Lamarck did not refute the possibility that microbes may form spontaneously from inanimate matter. As anyone who has worked with pure cultures of microbes can tell from experience, it requires considerable care to avoid contamination with unwanted creatures, and two centuries ago it was not appreciated how easily microbes get around. In other words, the contemporaries of Spallanzani were unable to reproduce his results and went on observing spontaneous generation. (In spite of this scepticism, the invention of canned food by Appert around 1800 was generally welcomed although this is in fact an applied version of Spallanzani's experiment.)

There was another reason why many were sceptical of Spallanzani's work. Towards the end of the 18th century studies on the chemistry of gases were initiated and Antoine Lavoisier had shown that access to oxygen was vital for animals. In Spallanzani's experiment all oxygen had been stripped from the infusion through boiling and this could then explain why the generation of microbes failed. The fact that many microbes not only tolerate, but even require the absence of oxygen was discovered only about half a century later. Theodor Schwann showed that if air was supplied to infusions via heated copper tubing, microbes did not appear. The objection to this experiment was that heating might destroy some 'vital principle' in air. Others then showed that if the air was passed through dry cotton wool this also prevented the entry of microbes—but still this could not convince everyone.

The experiments of Louis Pasteur around 1860 must be understood in the context of this background. Glass flasks with a drawn out, S-shaped opening and filled with sterile broth did not develop microbial biota. But if the neck of the flask was broken so that microbes could fall down into the contents of the flask, then bacteria soon took over. Pasteur, using a microscope, also demonstrated that air contains microbes. One would believe that the debate was now over. But as everyone knows, the British have a sceptic attitude towards continentals and the idea of spontaneous generation was defended in print in England as late as 1872. The physicist John Tyndall therefore carried out a series of experiments that finally showed that microbes do not originate spontaneously, but only from other microbes. Tyndall also showed that some bacterial spores survive boiling; this observation led to the invention of the autoclave for sterilizing material by exposure to water vapours at 120 °C under pressure.

In a certain sense ideas about spontaneous generation are still not quite dead. To be sure, no one today suggests that microbes can be generated in sterile hay infusions. But (pseudo) explanations for the origin of life due to 'inherent properties of matter', 'emergent properties', or 'self-organizing principles' are still common in the literature (especially authored by non-biologists)—and these speculations reflect a more or less continuous tradition going back to Aristotle.

The debate on spontaneous generation overlaps in time with Darwin. Darwin believed—as did Pasteur and Tyndall—that microbes are not formed spontaneously, but he added the word 'today'. This is because life must have originated from dead matter on at least one occasion. Darwin speculated that life could have originated in a 'warm little pond' where protein could have synthesized and potentially developed into life in a world where life was not already established. Today, however, if such a hypothetical proto-organism emerged it would not be observed because it would soon succumb to predation by already established organisms.

All of this shows a dilemma when discussing the origin of life. The insight that organisms always originate from other organisms and never from inanimate matter was hard won. But in a sense a spontaneous generation must have occurred at least once—and so parts of this book deal with spontaneous generation in the past.

The Panspermia hypothesis

The Swedish chemist Svante Arrhenius suggested in 1908 that life had originated somewhere else in the universe and in fact permeates all of Cosmos. According to this view the Earth had been colonized by microbes that had somehow escaped from their home planets and, driven by light pressure, had arrived on the Earth at some time in the past. This idea, the so-called Panspermia hypothesis, has haunted most of the 20th century. Supporters include such prominent persons as Francis Crick (who together with James Watson and others discovered the structure of DNA), although he later claimed that it was meant only as a provocation to show how difficult it is to understand the origin of life. The astronomer Fred Hoyle (also known as an author of science fiction novels) seemed, however, to have taken the Panspermia hypothesis seriously.

The following chapter discusses the possibility of extraterrestrial life. Here it should only be mentioned that the chance of finding life in our immediate cosmic neighbourhood, that is the solar system, is remote. But even if, against all odds, there should be life on other planets in the solar system, the Panspermia hypothesis has little appeal. It must be considered extremely unlikely that any microorganism could survive exposure to cosmic radiation that would result even from brief travel through space. The probabilities that life could survive transport on a meteorite or a comet that might have collided with Earth at some point in time is also extremely slight. Nor would DNA or RNA molecules survive cosmic travel in an intact form. To which should be added that transfer of such intact molecules, if possible at all, would not lead to anything. An isolated genome (e.g. a DNA molecule) that might have landed in a pool of organic molecules—or whatever one may have in mind—could not accomplish anything. A DNA molecule that is not a part of an intact living cell with specific enzymes that catalyse transcription and translation, a cell membrane, and without energy metabolism, is entirely incapable of creating life.

But the fanciful appeals to some people, and there is another reason why the Panspermia hypothesis has survived. If the origin of life is believed to be an extremely unlikely event, then relegating it to somewhere outside the solar system would seem to give more time. But in fact the age of the universe provides little help. To see this, we may imagine that the origin of life was caused by a random coupling of amino acids to form a functional enzyme—an enzyme that could, for example, catalyse the replication of a RNA molecule (that was also somehow synthesized by chance). The problem is similar to letting a rhesus monkey punch the keys of a typewriter. One day our primate will suddenly have written Shakespeare's *Hamlet*—it is only a matter of patience before this happens. Let us develop the idea in a somewhat more concrete form. Assume that the Earth's oceans consist of a solution of 20 different amino acids and that within each cubic centimetre of seawater one coupling between amino acids takes place every second. (We ignore for the moment that amino acids do not spontaneously condense to polypeptides in an aqueous solution.) We then await the formation a protein consisting of 100 amino acids coupled together in one particular sequence. The expected time for this to happen can be calculated to be 6×10^{51} years. The simplest organisms known, bacteria, contain about 1000 different proteins (and most are longer than 100 amino acids). The time available in our solar system is only about 4.6×10^{9} years and cosmologists now believe that the universe is about 1.5×10^{10} years old. The precise assumption of the example is not important here: it still robustly shows that a functional cell could not possibly have arisen spontaneously in a universe of finite age.

A final argument against the Panspermia hypothesis may appear unscientific: the idea is intellectually unsatisfactory. What is interesting is to understand *how* life could originate. It is very hard to make the Panspermia hypothesis appear probable—but it also represents only an attempt to export the problem to somewhere outside Earth without explaining how life could arise.

The Oparin–Haldane hypothesis—the primordial soup

During the first three decades of the 20th century the scope for understanding the origin and early development of life was very limited. The German zoologist Ernst Haeckel popularized the concept of 'primordial slime' (and indeed during the Challenger Expedition some sort of precipitate was recovered from the deep sea that was afterwards claimed to represent this hypothetical material). The primordial slime represented the idea that some sort of organic material had become alive and later gave rise to more advanced forms of life. In reality, the primordial slime hypothesis was just a variety of the idea of spontaneous generation. At the beginning of the century, many believed that the first organisms must have been photosynthetic algae since it seemed that only photosynthesis could generate organic matter from inorganic compounds. Thus all other kinds of life depended on green organisms.

In spite of great advances in biology and biochemistry during this period a more profound understanding of what life is, and especially the basic mechanisms of heredity, was first achieved with the development of molecular genetics in the 1940s. However, the discovery (or rather re-discovery) of genes as particulate units of heredity in 1900 did, early in the century, lead to the suggestion that life had originated as single genes, a view supported by, among others, the American geneticist Herman Muller. J. B. S. Haldane (see below) suggested virus as the progenitor of life. At the time, the understanding of the nature of genes and of viruses was still too incomplete for developing these ideas further. But they do represent an early recognition of the central problem: the origin of self-replicating units that are subject to Darwinian evolution.

A different approach was to try to understand the chemical and physical environment that might have prevailed on the young Earth and that led to the origin of life.

The Russian botanist A. I. Oparin and the English geneticist and physiologist J. B. S. Haldane suggested independently—in articles published in 1923 and 1929 respectively—ideas that still play a role in discussions of the prebiotic Earth and the abiotic synthesis of organic compounds as a prerequisite for the origin of life. Initially these ideas received little attention. This changed with the publication of Oparin's book (1936) and its translation into English (1938) under the title *Origin of life.*

Oparin and Haldane argued that organic molecules must have occurred before the origin of life and that the metabolism of the first organisms depended on the presence of such molecules in the environment. They further suggested that the primordial atmosphere was anoxic and chemically reducing (crudely speaking that it contained more hydrogen than oxygen). The atmospheric composition would include compounds such as methane (CH_4), ammonia (NH_3), hydrogen sulphide (H_2S), and possibly hydrogen (H_2) and carbon monoxide (CO), in addition to water (H_2O) and nitrogen (N_2). That elemental oxygen was practically absent in Earth's earliest atmosphere is today considered a fact; how chemically reducing it was is still a matter of debate (see Chapter 14). There were several reasons to assume an early oxygen-free atmosphere. First of all, free oxygen is an unlikely constituent of a planetary atmosphere since, if present, it would quickly combine with hydrogen, iron, sulphur, and carbon, and O_2 does not occur in the atmosphere of other planets. The composition of the planet Jupiter's atmosphere (largely CH_4, H_2, and NH_3) was already known when Oparin wrote his book. In fact a primordial anoxic atmosphere of the Earth had been suggested earlier by Arrhenius and by the Russian geochemist Vernadsky, and they understood that the presence of atmospheric O_2 in the extant atmosphere is a result of photosynthesis of green organisms. There was an additional reason for Haldane and Oparin to assume an O_2-free atmosphere: organic compounds are not stable in the presence of oxygen since they would eventually become oxidized to form principally CO_2 and H_2O. Finally Oparin suggested a number of mechanisms and synthesis pathways by which a number of simple, but biologically important compounds such as amino acids, volatile fatty acids, and carbohydrates could have formed from abiological processes in the oceans and atmosphere of the young Earth.

The next step in the reasoning of Haldane and Oparin was that the energy metabolism of the first organisms depended on fermentation. There are several types of fermentation (see Chapter 7), but they have in common that energy is provided from splitting larger organic molecules into smaller ones. Fermentation does not require an external oxidant (electron acceptor) and organisms with a fermentative energy metabolism are therefore independent of O_2 in the environment: they are anaerobes. Well-known fermentation processes include lactic acid fermentation (the degradation of glucose molecules into two molecules of lactic acid), ethanol fermentation (in which glucose is broken down to ethanol + CO_2), and butyric acid fermentation (decomposition of carbohydrates into butyric acid + H_2 + CO_2). At the time it was also assumed that fermentative metabolism is relatively simple and its biochemistry was to some extent already understood by then, whereas much less was known about respiration and phototrophy (light-dependent energy generation). Respiratory carbohydrate catabolism is initiated by an anaerobic fermentation (the glycolytic pathway) in, for example, animals. Oparin assumed that this represents a relic from a time when energy metabolism was only fermentative. Today, most will probably disagree on this point and assume that a primitive respiratory or phototrophic energy generation came earlier and that fermentation is a later development; in fact, fermentative processes are not mechanistically simple (see Chapter 7).

In 1953 Stanley L. Miller attempted to make a primordial atmosphere (in the first experiment consisting of CH_4, NH_3, H_2, and H_2O) that was contained in an apparatus as shown in Figure 3.1. The expected synthesis of organic compounds would require energy that was supplied in the form of electric sparks (a sort of Precambrian lightning). After a week the liquid was analysed. In addition to some tar-like polymers with an unknown chemical composition, several biologically important monomers were detected including some amino acids.

This experiment marked the initiation of a new research field known as *prebiotic chemistry*. In later experiments the chemical composition was varied using, for example, CO or CO_2 rather than CH_4 as carbon source (thus considering a somewhat less reducing primordial atmosphere). It has also been tried with H_2S and other more or less plausible simple compounds, and with non-specific catalysts such as various metal ions. Other energy sources have also been used such as ultraviolet radiation. Since the early atmosphere did not contain O_2—and therefore no ozone layer—the intensity of UV radiation must have been considerably higher at the surface of Earth than is the case today.

In this way it has been possible to observe the synthesis and synthetic pathways (many of which were already known by chemists) for many molecules that are central to life. Thus all of the 20 essential amino acids (plus several others that do not

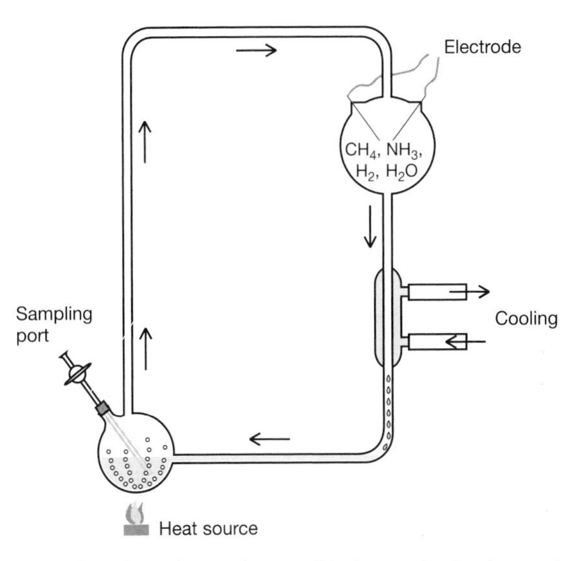

Fig. 3.1 Stanley L. Miller's (1953) experiment with the synthesis of organic compounds in a 'prebiotic atmosphere'.

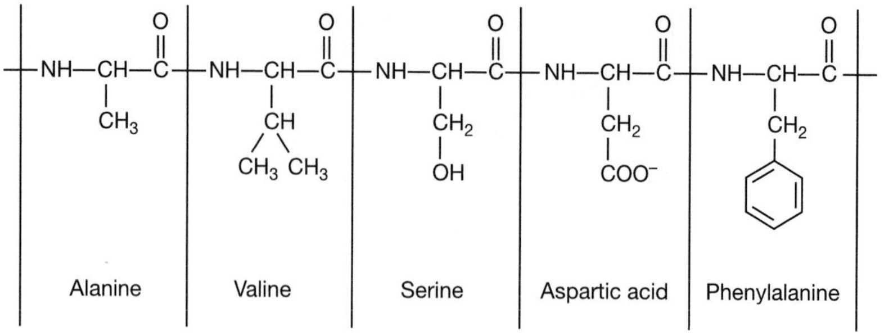

Fig. 3.2 Proteins (polypeptides) are linear chains of amino acids of which five different ones are shown. They are coupled together by splitting off a water molecule. Redrawn from Miller and Orgel (1974).

occur in organisms) have been produced in such experiments (Figure 3.2). A possible pathway for amino acid synthesis is via the reaction between aldehydes and cyanide. It is striking that those amino acids that are produced in highest amounts in such experiments (e.g. alanine, glycine, leucine) are also the quantitatively most important constituents of proteins. Volatile fatty acids (formate, acetate, propionate, butyrate) and urea are also easily synthesized in Miller-type experiments. Porphyrins, representing the basic molecular structure in important biomolecules such as respiratory enzymes, chlorophylls, and haemoglobins (see Figure 7.2) have also been synthesized under certain conditions.

Synthesis of the five organic bases *adenine*, *guanine*, *cytosine*, and *uracil* (in RNA) or *thymidine* (in DNA) is, of course, of special interest since they constitute the 'letters' of the genetic code. All have been synthesized in Miller-type experiments under more or less plausible prebiotic conditions (adenine, for example, can be formed as a polymer of 6 HCN). However, in RNA and in DNA these bases are bound to a carbohydrate with five carbon atoms (ribose in RNA, deoxyribose in DNA). Such a unit consisting of one of the organic bases and a carbohydrate molecule is called a *nucleoside*. If further a phosphate group is added the unit is called a *nucleotide* and that is the building block of nucleic acids (Figure 3.3). Prebiotic chemistry has not been successful in synthesizing the carbohydrate components. Carbohydrates can be synthesized (by condensation of formaldehyde molecules), but as in many other chemical processes that are not controlled by specific catalysts, many side-reactions and by-products mean that the desired molecular species are produced in extremely small quantities. Furthermore, it has proven difficult to join the carbohydrate, the phosphate, and the bases correctly under putative prebiotic conditions. The inability to explain how nucleosides can be synthesized under plausible prebiotic conditions remains a serious obstacle when attempting to understand the origin of life.

Another disappointment has been a poor yield of longer fatty acids that, through condensation with glycerol, are necessary for the synthesis of lipids. Simple ideas about the origin of the cell membrane are thus not easy to support.

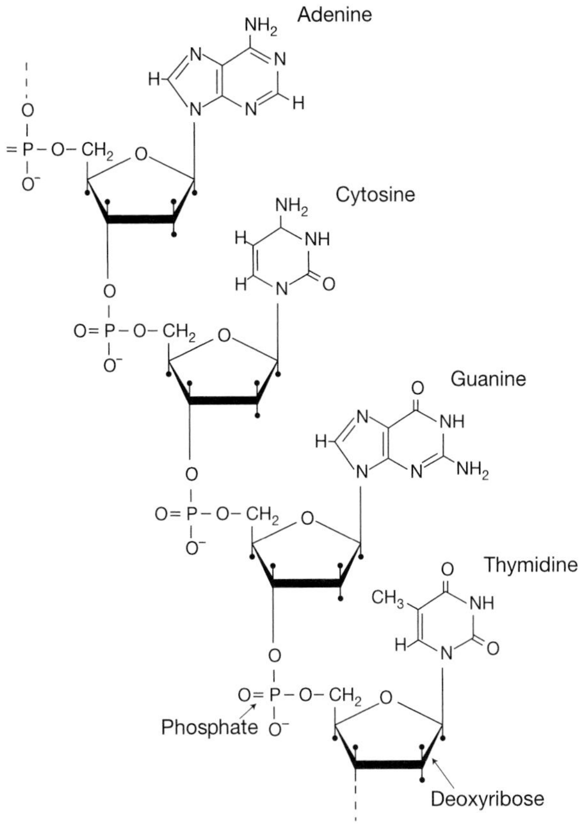

Fig. 3.3 Structure of a DNA strand. It consists of sequences of four different organic bases held together by a backbone of phosphate and deoxyribose molecules. Redrawn from Miller and Orgel (1974).

Miller-type experiments may yield fatty acids, but they are branched in contrast to the linear molecules of living systems.

At this point it should be stressed that the Oparin–Haldane hypothesis and its experimental support do *not* represent any explanation for the origin of life. At the best they present a picture of the plausible chemical environment in which life originated and they do explain the origin of a number of central and universally occurring molecules in living organisms. The idea about this chemical starting point for life has been popularized through the concept of the *primordial soup* (probably originally from Haldane's description of the prebiotic ocean as *a hot thin soup*). Underlying is the view of the primordial oceans as a sort of broth. The concentration and composition of this soup is, however, quite open for speculation. The organic molecules that formed were not stable and would eventually degrade (e.g. by short wavelength irradiation) and some sort of equilibrium—that we cannot estimate—must have been approached. Even so, the Oparin–Haldane world has provided an important contribution for understanding the origin of a number of central chemical constituents of life. It seems to explain why particular molecular structures (such as the purine and pyrimidine bases, the essential amino acids, and porphyrin), rather than a multitude of other possible groups of organic compounds, are of universal importance in living organisms.

The synthesis of organic molecules in Miller-type experiments is not only a laboratory phenomenon. This is evident from the fact that organic compounds (hydrocarbons, various amino acids) have been found in chondritic meteorites and especially in the Murchinson meteorite that hit Australia in 1969. These results were in periods controversial since amino acids of biological origin are omnipresent on the surface of Earth. Today it seems generally accepted that the findings did not result from contamination and that the meteorite did contain amino acids. The fact that non-biological amino acids as well as both D- and L-forms (see below) occurred, support this. Simple organic molecules have been detected by spectrometry in the atmospheres of the giant planets, in the atmosphere of the Saturn moon Titan, in comets, and in interstellar dust clouds. The abiological synthesis of organic molecules is therefore a widespread phenomenon in the universe. It has even been suggested that the transport of organic compounds to the prebiotic Earth via collision with comets and meteorites played a role in the origin of life.

We have already mentioned some disappointments with experimental tests of the Oparin–Haldane hypothesis. They also fail to elucidate some other central questions. These include the formation of polymers (e.g. polypeptides from amino acids and nucleic acids from nucleotides), the central role of certain metals in many enzymes, and the universal role of phosphate in biological energy transfer. In this connection, various other hypotheses have been suggested and among them especially, ideas about the synthesis of organic compounds on mineral surfaces. Some of these ideas have been further developed to explain a primitive energy metabolism and even a sort of genetic system, but we will return to these aspects later. The possible mechanisms described below are not necessarily an alternative to the Oparin–Haldane world, but may also be understood as being complimentary to it.

Other models for prebiotic chemistry

Other models of prebiotic chemistry all show one weakness relative to the Oparin–Haldane model in that they have only to a very limited extent been experimentally supported. A hypothesis suggested by Cairns-Smith is based on the synthesis of organic compounds on the surface of clay minerals. Clay minerals are crystalline and the hypothesis includes the idea that heredity was originally associated with crystal surfaces ('crystal genes') in that variations or irregularities of the crystal lattice are maintained (inherited) as the crystal grows. One aspect that can be explained through Cairn-Smith's ideas is the polymerization of amino acids. When amino acids combine to form polypeptides, this is a condensation process resulting in loss of a water molecule. In aqueous solution the equilibrium

$$\text{monomers} \leftrightarrow \text{polymer} + H_2O$$

is shifted strongly towards the left side of the equation. Adsorption of amino acids to clay surfaces should shift the equilibrium towards the right side of the equation due to the higher local monomer concentration. This effect has actually been demonstrated experimentally. Other ideas, in the context of the Oparin–Haldane hypothesis, have been based on polymerization of amino acids due to evaporation in coastal pools, or increased concentrations in the liquid phase during formation of sea ice.

An important problem that must be addressed in any complete theory for the origin of life is that amino acids from natural proteins always show the same *chirality*. Complex three-dimensional molecules, such as carbohydrates and most amino acids, occur in two forms: a left- and a right-form, like the right and left hand; these are referred to as L- and D-amino acids. Natural proteins are built exclusively from L-amino acids and RNA and DNA contain only D-carbohydrates. Miller-type synthesis always yields an equal number of L- and D-amino acids. If polypeptides originated on crystal surfaces it is conceivable that there is a preference for a given chirality. The clay mineral hypothesis may also explain the central role of phosphate in cellular energy transduction. Clay minerals have a high affinity for dissolved, negatively charged ions such as phosphate, and if life originated on clay surfaces the important role of phosphate could perhaps be explained.

Pyrite ('fool's gold', FeS_2) has been advocated by G. Wächterhäuser as the original site of organic synthesis and also for the origin of life. Pyrite is a mineral that is formed under anaerobic conditions, for example in marine sediments that are rich in sulphide (derived from bacterial sulphate respiration) and iron (which in one form or another is ubiquitous on the surface of Earth). Pyrite is also formed in hot, sulphidic springs. Large pyrite crystals shine like brass and are often sold in kitsch shops as decorative objects. Pyrite can be formed through the process:

$$Fe^{2+} + 2HS \rightarrow FeS_2 + H_2.$$

The hydrogen produced by this process could thus, according to Wächterhäuser, reduce CO_2 to organic compounds. The synthesis of organic compounds from CO_2 coupled to the reaction between Fe and Ni with sulphide has recently been demonstrated experimentally. The hypothesis nicely explains the universal role of Fe-S groups in enzymes involved in electron transfer in cellular processes (e.g. respiration). The role of other metals in various enzymes and the role of phosphate in metabolism may perhaps also be understood in the light of the Wächterhäuser hypothesis, and it has been extended to explain the origin of phototransduction as a forerunner of photosynthesis. The model is in many respects attractive in terms of the evolution of energy metabolism, but its weakness is a very limited experimental verification.

Some other, more or less related ideas on prebiotic chemistry have been suggested; among them some have a certain appeal. But like hypotheses based on crystal surfaces, they suffer (perhaps to an even greater extent) from meagre experimental support.

Coacervates

Oparin worked over a longer period with a special property of certain colloids and mixtures of colloids (aqueous solutions of very large molecules, egg white and wallpaper paste may serve as examples). These mixtures may form small (1–500 μm) spheres with a certain (superficial) likeness to living cells. Oparin called these spheres *coacervates*. Not all types of colloids form stable coacervates. The interesting ones form a kind of spherical membrane surrounding an aqueous interior. Materials that have been used include protein + gum-arabic and albumin + RNA. Sidney W. Fox (USA) has more recently continued the experimental study of coacervates.

One could say that preoccupation with these amoeba look-alikes is naïve and could hardly contribute much insight into the origin of life. However, coacervates have interesting properties that may throw some light on the origin of energy metabolism and cell division.

Dissolved enzymes have a tendency to become trapped inside coacervates during their formation. In one experiment, the enzyme phosphorylase accumulated inside coacervates (Figure 3.4). When glucose-1-phosphate was added to the water the small molecules could diffuse inside the coacervates. The phosphate bond then delivers the energy that drives the following processes. The enzyme releases the phosphate and at the same time polymerizes the glucose units to starch. The released phosphate can diffuse out through the walls of the coacervates, but the large starch molecules are trapped inside. The coacervates therefore grow and eventually divide into two coacervates. Since the enzyme thus becomes diluted inside the coacervates the processes eventually come to a halt.

In another experiment, the redox enzyme NADH dehydrogenase was trapped inside coacervates. When the redox dye methyl red and NADH (reduced nicotinamide adenine nucleotide) are added to the water both compounds can diffuse into coacervates. Here the enzyme catalyses the reduction of methyl red coupled to the oxidation of NADH to NAD. This represents an analogy to a simple electron transfer system.

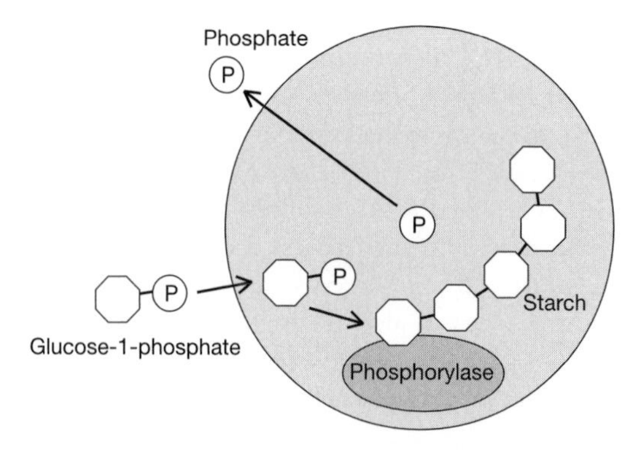

Fig. 3.4 One of Oparin's experiments with coacervates. The enzyme phosphorylase and the starch molecules being formed are trapped within the coacervate, whereas glucose-1-phosphate and phosphate can pass through the 'cell membrane' of the coacervate.

One interesting property is therefore that coacervates can model the spatial organization of simple bioenergetic processes. Another one is that, for purely physical reasons, coacervates tend to maintain a certain size and they therefore divide if their volume increases. A third interesting aspect is that such structures, if they could arise in a prebiotic world, could possibly have provided spatial structures in which the first replicators could reside. As explained in Chapter 6 such a spatial structuring or compartmentalization appears to be a necessary prerequisite for the earliest development of life. Spherical vesicles based on a double layer of lipids are perhaps more interesting as a model for a primitive cell membrane; and this is discussed later in this book. Some more recent theoretical approaches to the origin of life (notably by Freeman Dyson) are directly inspired by Oparin's work and assume that coacervates of some kind necessarily preceded genetic systems. These ideas are also discussed in more detail in Chapter 6.

Chapter 4

What is life?

Equipment in a space probe—exobiology

Something with eyes or antennae that crawls or walks is easily recognized as 'life'. When shown a dried seed or a slimy bacterial film, many people would be less certain. In fact, it is not easy to define life. Among many other difficulties it is a problem that we know of only one kind of life.

It is possible to rephrase the question by asking what sort of scientific equipment should be placed in a space probe that is sent off into space to investigate whether there is life on some other planet. This was topical when planning the Viking mission that landed two probes on Mars in 1976 and again today in connection with future Mars missions. In the case of the Viking probes, Martian soil was analysed for organic matter (which as we have seen does not prove the presence of life). Furthermore, radioactively labelled organic substrates were added to Martian soil to see whether this was degraded into CO_2, and labelled CO_2 was added to see whether it would become assimilated into organic matter. If the probe had landed on Earth, the tests would have been positive almost any place. On Mars, as is widely known, these tests proved negative.

One might think that these experiments were naïve by assuming that if there were life on Mars it would behave exactly as life as we know it on Earth. Life on Mars could be based on quite different types of metabolic processes that might not occur on Earth; also many kinds of microorganisms known from Earth would not, in fact, respond positively to these tests. It is very difficult to decide what to measure in order to document the presence of life—even if there were no practical or economic limitations to what sort of equipment could be put into a space probe. One suggestion would be to test for a composition of the local atmosphere that was far from chemical equilibrium. Thus the oxygen-rich atmosphere of the Earth would seem to be very improbable in the absence of (biologically-mediated) oxygenic photosynthesis. As we will see later, an environmental chemical disequilibrium and element cycling are necessary, but not sufficient conditions for the presence of life. Even if Earth had remained lifeless its surface would not have been in complete chemical equilibrium.

Before we attempt a more detailed definition of life it may be relevant to discuss the possibility of extraterrestrial life. The study of extraterrestrial life has given rise to a new discipline referred to as *exobiology* (or sometimes as *astrobiology*). As such it is highly unusual insofar as it is unknown whether its subject matter exists. It is perhaps also peculiar in that biologists are among those who are least enthusiastic about this speculative topic.

It is possible to constrain the limits of the physical environment that allows for the existence of life, even when we stretch our imagination to the utmost. To the extent that possible extraterrestrial life in any way resembles that on Earth, then water in a liquid form is an absolute requirement. This implies that active life is limited to a narrow combination of temperature and pressure. At the surface of Earth (with an atmospheric pressure of about 1 bar) this means from slightly below 0 °C and up to 100 °C. There are actually bacteria that may grow below 0 °C whereas others manage up to about 85 °C. The latter, so-called 'extreme thermophilic bacteria', may at higher pressures (such as in cracks in rocks about 1 km beneath the surface, or around hydrothermal vents in the deep sea) grow above 100 °C, but a temperature of 115 °C appears to be an absolute limit. This collective span in temperature tolerance depends on the evolution of special enzymes with different temperature stability properties and lipids with different melting points in membranes. We are here speaking only about prokaryotes: the collective span in temperature tolerance of eukaryotes is only about 0–55 °C.

Now one might imagine life that does not require water as a universal solvent, and is not based on carbon chemistry, but perhaps depends on complicated silicon chemistry. But even so, a relatively narrow temperature interval would be indicated that could not deviate much from what is found on the surface of Earth. Even at moderately low temperatures, all possible solvents become solids and the thermal motion of molecules would play a modest role. If life is to be imagined under such conditions it would have to be in the form of mechanical devices based on pendulums, gears, etc. Conversely, at higher temperatures, where molecules have a high kinetic energy, compounds melt and evaporate and large complex molecules split into simpler ones and eventually into atoms. Life as we know it is in part based on quasi-crystalline structures (genetic information in the form of RNA and DNA, cell membranes with their complex structures). But life processes also depend on the transport of dissolved molecules by molecular diffusion and on chemical reactions in solution. Life is thus in a subtle way built from both solids and liquids. And so temperatures that allow for life cannot deviate strongly from those of the Earth's climate.

Life must have access to free energy in order to maintain integrity and growth. On Earth the all-dominating energy source is electromagnetic radiation from the sun. The wavelengths used range between 400 and 1000 nanometres (visible and near infrared light, the latter being used by some photosynthetic bacteria). Photosynthetic organisms convert this energy source into a chemical potential (primarily as organic matter here symbolized by $[CH_2O] + O_2$) and this chemical potential fuels the rest of the biosphere. (A mixture of oxygen and hydrogen gases is an example of a 'chemical potential'—applying a lit match will liberate the energy.) It is not an accident that life utilizes only this narrow span in wavelengths of electromagnetic radiation. Light with longer wavelengths consists of light quanta with an insufficient energy to liberate electrons from molecules and thus accomplish chemical reactions, and its absorption results only in heat. Electromagnetic radiation with shorter wavelengths, on the other hand, consists of much more energetic quanta leading to an unspecific ionization of molecules and the resulting reactions would be difficult to control.

Radioactive decay in the Earth's mantle and crust contributes, through heat generation, to creating a chemical potential ($H_2 + CO_2$ or $H_2 + SO_4^{2-}$) that is utilized in particular by methanogenic and by sulphate reducing bacteria—as seen in geothermal springs and deeper in the crust of the Earth. This is quantitatively of limited importance and from an evolutionary viewpoint it probably represents a secondary adaptation: hydrogen gas is produced in copious amounts in fermentative processes in anoxic habitats such as marine and limnic sediments and so methanogenic bacteria and sulphate reducers are widespread in less exotic habitats.

Within the solar system, the presence of life is impossible on the inner planets Mercury and Venus due to the extreme temperatures and absence of water. Mars is so far the most likely possibility, although prevailing conditions do not appear promising. The thin atmosphere (0.5–1% of atmospheric pressure on Earth) consists mainly of CO_2 and with an extremely low water pressure corresponding to the vapour pressure of ice at temperatures below −80 °C. Mars' poles are covered by frozen CO_2, but they also contain some water ice. The red surface is due to oxidized iron. The low water content reflects loss of hydrogen through the history of the planet and its surface is oxidized throughout. It has been claimed that some terrestrial life could *survive* on Mars (indeed many microbes can survive in liquid nitrogen), but this does not mean that they can grow there. The virtual absence of water, the constantly low temperatures, and absence of potential chemical (reducing) substrates completely exclude the possibility that any form of life as we know it could thrive on the surface of Mars. It has also been suggested that life may occur beneath the surface in cracks in rocks—just as metabolically active bacteria have been found on Earth down to a depth of at least 1 km. As already mentioned, on Earth this is based on radioactive decay; it cannot yet be entirely refuted that such conditions (including liquid water) occurs on Mars and this will probably be studied in connection with future Mars probes.

Structures resembling dried river beds and deltas on Mars' surface have led to the idea that it once had free water, a denser atmosphere, and active volcanism that allowed for a more temperate climate and chemically more reducing conditions. That is, that the Earth and Mars resembled each other more in the early history of the solar system. If so, it is not impossible that life did once arise on Mars, and it is even possible that some evidence for this may be found. It has also recently been suggested, however, that the surface structures on Mars were caused by explosive evaporation of buried, solid CO_2. The implication is that Mars was never wet and that it always had a thin atmosphere. The somewhat sensationalized report about bacterial fossils in a meteorite that with considerable probability derives from Mars' surface has not convinced the majority of biologists and palaeontologists about the existence of life on Mars in the past.

Among other components of the solar system there has been some interest in the Saturn's cloud-covered moon Titan. It has a nitrogen atmosphere containing methane and some organic compounds have also been detected in its atmosphere. Possibly there are lakes or seas of liquid methane. It has been suggested that Titan is in a 'primordial soup stage' like the one Earth is supposed to have passed through early in its history. Jupiter's moon Europa is ice-covered and circumstantial evidence suggests that there is liquid water beneath the ice, which is kept warm by energy generated by the tidal forces of Jupiter. Europa represents the last theoretical possibility for extraterrestrial life in the solar system. Both in the case of Europa and Titan the low level of electromagnetic radiation that reaches these celestial bodies renders it unlikely that life could have arisen and be maintained there. Technically and economically it is probably possible to land space probes on both moons.

When it comes to the universe beyond the solar system one could, by sheer statistical reasoning, assume that there is life somewhere. The Milky Way contains about 10^{11} stars and there are at least 10^{10} other galaxies within the visible part of the universe. And so it is reasonable to think that somewhere there will be planets with physical and chemical properties resembling those on Earth about four billion years ago—even though it is still entirely unknown whether this alone constitutes sufficient conditions for the origin of life, or whether other very unlikely events were also required. But under all circumstances there are many constraints. Stars that are larger than the sun have a short life span and cannot be taken into consideration and stars that are somewhat smaller maintain too low a radiation intensity for planets to receive sufficient energy. This leaves perhaps 10% of the stars. It is now known that there are stars other than the sun that have planets, but it is not known how frequent this is, let alone the frequency of planets with sizes comparable with Earth. It is obvious that with respect to planet sizes and distances to the mother star there are constraints. Small planets rapidly lose their atmosphere (Mercury, the moon) and large planets become gas giants, such as Jupiter. Even planets with sizes similar to the Earth may develop an extreme climate (Venus).

A further condition for life as we know it is the availability of several chemical elements. There are two reasons for this. First of all, all known organisms require at least 24 different elements as essential components (see the following section). The other reason is that, if not the *origin* of life, then at least its maintenance requires that the Earth is geologically active. Perhaps one may consider volcanism and earthquakes as natural catastrophes. But without volcanic out-gassing and without mountain building and the accompanying erosion, then essential components such as CO_2 and soluble phosphate would accumulate in sedimentary rocks and there would be no mechanism to bring them back to the biosphere. Life would thus run out of essential building blocks and eventually vanish. The energy source that drives geological processes is radioactive decay in the Earth's crust and mantle, and this is especially due to isotopes of uranium, thorium, and potassium. Originally the universe contained only hydrogen, helium, and traces of lithium—and this does not give rise to very interesting chemistry. The remaining elements formed later through fusion processes in the interior of stars. The oldest stars therefore contained only hydrogen and helium whereas younger stars contain heavier elements in their surface and on their putative planets because debris from older stars contributed to the formation of these younger stars. In ordinary stars, however, only elements that are lighter than iron form—and many heavier elements such as copper, molybdenum, and zinc are biologically essential. Such elements form only in connection with super nova explosions—the cataclysmic end of existence for very large stars. The reason that the solar system—and thus Earth—contains relatively high amounts of heavy elements is that material from an earlier super nova explosion (that must have taken place in the vicinity of the region of the Milky Way where the sun originated) were incorporated into the forming solar system. The Earth's content of naturally radioactive elements represents part of the energy that was released during this super nova explosion earlier in the history of the universe. Many stars (and their hypothetical planets) represent a simpler chemical composition and are unlikely to harbour life.

In other words, whereas statistical considerations may suggest that life occurs elsewhere in the universe, this is certainly not a frequent phenomenon. Some have attempted to calculate probabilities for this, but in my view this is waste of time, and under all circumstances we will never be able to know! The two closest stars outside the solar system are respectively about four and eight light years away—and even from these small distances (in a cosmic scale) we will never be able to receive information from a space probe.

Undeniably it would be interesting to study life that had developed independently from what we know on Earth—and even though the odds are very poor, it seems worthwhile to look for it in the few possible sites within the solar system. But when grown-up people try to listen for 'intelligent' radio signals from the universe, this does seem to be waste of time and resources (in the USA, for a period, even tax payers' money). The probability of 'intelligent' (in reality meaning 'technological') life existing elsewhere is infinitely low. Even on Earth the origin of human beings is an accidental result of Darwinian evolution that would not be likely to recur if time could be turned backwards to an earlier geological period. Also, it should be kept in mind that the time period during which there has been so-called intelligence on Earth (that is, use of radio signals for communication—the 'post-Marconi period') is only about 10^{-8} of the period during which there has been life on Earth. If this number is multiplied with some equally low probability for extraterrestrial life in any form we will end up with a very small number indeed. Also, we would not be able to receive radio signals transmitted from even the closest star if our cosmic friends use similar transmission power as we do.

Life—chemical composition

Life is a kind of complicated chemistry, so it is natural to start discussing its chemical composition. There are altogether 92 naturally occurring elements on Earth. Among them more than one-quarter are vital for all kinds of life, and these elements are relatively common in the Earth's crust and in seawater. Apart from water, organisms are essentially built up from organic compounds and especially proteins, nucleic acids, lipids, and carbohydrates. These are all compounds consisting of C, H, and O with a variable content of N, P, and S. These six primary elements each constitute more than 1% (O up to 60% and C 5–10%) on a weight basis. Oxygen is so important in terms of weight because water usually exceeds 60% of organism weight. The essential electrolytes: Na, K, Mg, Ca, Cl each represent between 0.05 and 1% of the weight of cells. All the other essential elements, although of vital importance, each constitute less than 0.05%. Several metals: Fe, Mn, Cu, Co, Mo, Ni, Zn, W, and V, are necessary components of various enzymes and I, F, Br, and Se are also included among the essential elements. Si, Ba, and Sr are important constituents in cell walls and skeletons of certain organisms, but it is unknown whether they have any further functions. A number of elements, such as As, B, Cd, Cr, Hg, Pb, and Li, are to a variable extent accumulated by different organisms, but whether they play a functional role is unknown. It is often difficult to establish whether trace elements are essential to organisms. It is easy to demonstrate the necessity of, for example, Fe and Cu, and precise functions of these elements are known. But trace elements could be essential in such small concentrations that it would be impossible to exclude them in experimental situations and enzymes containing them may still be undiscovered.

While all these elements are present on the surface of Earth, their concentration differs between organisms, soil, and seawater. Silicon, for example, is the next common element in the crust (after oxygen), but it plays a rather marginal role in organisms. Aluminium, another common element, does not as far as is known, play any biological role. Conversely, molybdenum and cobalt are essential elements that occur rather sparsely in soil and in seawater. A number of elements are probably of no importance to life; these include the actinides (U, Th), the precious metals, the rare earth elements, and the noble gases (He, Ar, Ne, etc.).

In the beginning of the 19th century organic matter was considered to represent special 'vital material' that was outside the reach of established chemistry. It was then found that organic chemistry is simply the chemistry of carbon compounds. Thereafter it was quickly learned how to synthesize simple organic molecules, and how to account for some simple biochemical processes.

At first sight the organic compounds from which life is built represent an overwhelming diversity. This diversity is first of all due to the proteins with their different and very specific catalytic or other properties, and even the simplest bacteria synthesize about 1000 different kinds. But proteins are just linear chains (polymers) of only 20 different kinds of amino acids and the diversity is simply a function of the sequence in which the amino acids are bound together. Other important organic molecules are also polymers of a limited number of rather simple molecules: DNA, RNA, and carbohydrates such as starch and cellulose. Irrespective of whether the organic molecule structures constituting life are components of polymers or function as individual molecules, they occur in practically all organisms and their number is surprisingly small. Apart from the 20 amino acids, there are a number of vitamins and coenzymes (e.g. biotin, porphyrins, nicotine dinucleotide, riboflavin, and some more); in cells these are often bound to particular proteins and may contain metal ions. Furthermore there are the pyrimidine and purine bases (altogether five that are the essential constituents of DNA and RNA), a handful of carbohydrates (e.g. glucose and ribose), and lipids (fatty acids, isoprenoids, occurring as glycerol esters). Some more could be mentioned including various intermediate metabolic products and some more special compounds occurring only in certain groups of organisms. But altogether, living organisms presumably contain less than about a hundred basic molecular structures with relatively modest variation. This universality as well as the limited number of molecular structures is astonishing. It should be recalled that even if we consider only organic compounds with relatively few carbon atoms, the number of possible combinations of C, O, and H seem almost endless.

The *CRC handbook of chemistry and physics* (78th edn, 1998) lists about 12 000 organic compounds—most of which exist only in chemical laboratories or are used in some technical or medical context, but do not occur in nature. And these 12 000 compounds constitute only a fraction of the possible CHO combinations. Biochemistry is thus only a small corner of organic chemistry.

The reason for this could, perhaps, be found in the kind of compounds that are synthesized in Miller-type experiments and may have been dominating in the prebiotic world. It does suggest that all basic biochemistry was established very early in the history of life—a theme that we will revisit later in the context of the origin of genetic systems and energy metabolism.

A couple of other properties of life

Life is cellular—the basic unit of life is the cell. Multicellular organisms like humans are colonies of cells. In this case it is sometimes debatable what is inside and what is outside—consider for example the intestinal lumen. Multicellular organisms also often include non-living material that may derive from dead tissue or from secretions (skeletal structures like cartilage, mollusc shells, or large parts of woody tissue). But as far as the basic unit—the living cell—is concerned, whether a bacterium, a protozoon, or a part of a colonial or multicellular organism, there is always a well defined inside and outside delimited by a cell membrane. To be sure, some cells have invaginations of the external cell membrane, but fundamentally the living cell is spatially well defined. The interior chemical environment is maintained within narrow limits even in a varying external environment—up to a point where the cell can no longer regulate its internal environment and it dies. The internal environment is maintained through complex interactions of feedback loops that, for example, control the uptake or release of different ions in order to maintain relatively constant internal ion concentrations and pH.

As we have seen, metabolically active cells consist mainly (but to a somewhat varying degree) of water. In a number of ways the H_2O molecule possesses unique properties, including its unusual efficiency as a solvent. It is not possible to imagine biochemical processes in the absence of water. Nevertheless, some cells (e.g. bacterial endospores, certain types of protozoon and algal cysts, and even some small invertebrates such as some rotifers and tardigrades) may survive total dehydration. But during dehydration all metabolic processes come to a halt.

A related, but perhaps more illuminating phenomenon, is that cells in principle may survive freezing down to temperatures approaching absolute zero (0 °K). In practice, ice crystals may destroy cells during freezing or subsequent thawing. But in many cases, using different precautions and in particular very rapid cooling, cell destruction can be avoided. It is now customary to maintain culture collections of microbes in liquid nitrogen (about −196 °C) and these can be kept viable for years. At such low temperatures all metabolic activity stops. But this also applies to the thermal processes that would otherwise, in accordance with the second thermodynamic law, degrade structures in the absence of energy metabolism (see the following section). Thus structures are preserved in the frozen cells, but these will not show any of the properties that could define life: they can be considered only as mechanical objects. But after thawing, normal life processes (energy metabolism, growth, and reproduction) re-start as if nothing had happened. Of course this is a well-known phenomenon, but it also shows a fundamental property of life: *function and information is solely associated with structure.*

Collectively, living things represent an enormous size span: small bacteria measure about 0.5 μm and a large whale perhaps 10 m, a range that exceeds seven orders of magnitude with respect to length or almost 22 orders of magnitude with respect to mass. We know of no other category of things, whether artefacts or natural objects, with such a size span, and it is astonishing that we can find common properties for bacteria and whales. However, and as already mentioned, the unit of living things is the cell, and whales are just colonies of an enormous number of cells. In our context it would be more interesting to discuss constraints on cell size.

The smallest metabolically active bacteria measure about 0.5 μm or about 0.06 μm^3. Is this a fundamental limit for minimum size? It is obvious that the cell must contain certain organelles in order to function. This includes a chromosome that in free-living bacteria carries at least 1000 genes corresponding to about 10^6 nucleotide pairs and if stretched out measures about 1 mm. The cell must also contain ribosomes, a cell membrane, and different macromolecules that all take up space. A more general minimum size limit derives from the fact that metabolic processes depend on molecular diffusion—and for diffusion along a concentration gradient to be a precise and predictable process a minimum number of molecules are needed (see further in the following section). As an example we may consider the necessary number of protons in cells. The function of energy metabolism is primarily that of a pump that transports protons out of the cell, thus creating an electrochemical gradient across the cell membrane (Chapter 7). The return flux of H^+ is then coupled to the synthesis of ATP that again is used for all energy-requiring processes in the cell. It takes a flux of $3H^+$ to generate one molecule of ATP. There is therefore a rapid turnover of intracellular protons. What is the necessary minimum number of cellular protons? We may first note that the pH at which cellular processes take place is about 7; this means that the H^+ concentration is 10^{-7} or that there are 10^7 H_2O molecules for each H^+ (pH is defined as the negative logarithm of H^+ concentration). One would think that, for example, 10 protons would be too few since this, among other things, would lead to unacceptable fluctuations in intracellular pH. If there were on the average 10 H^+ this would correspond to 10^8 H_2O molecules. One mole corresponds to about 6×10^{23} molecules and, in the case of water, to 18 gram or 18 cm^3. The water content of the bacterium would then be about 3×10^{-15} cm^3 or 0.003 μm^3. If the cell is spherical and largely consists of water this would correspond to a diameter of about 0.18 μm. This is smaller than found in real bacteria. If we, somewhat more realistically, assume that the intracellular contents is 100 H^+ then the minimum volume of water would be 0.03 μm^3 and the diameter about 0.4 μm, that is, approaching the size of the smallest real bacteria. It should be mentioned that there exist two categories of smaller bacteria. Starving and metabolically inactive bacteria have a lower water content and dispense with some of their organelles and so they shrink relative to active cells. The existence of 0.2–0.3 μm large so-called nanobacteria has recently been claimed—although this is still controversial. They probably represent such inactive cells. Also, bacteria that are intracellular parasites can dispense of a number of functions and structures and can depend on the host cell for a stable chemical environment. Generally, however, for a metabolically active, free-living cell—and thus for the basic unit of life—a minimum size of about 0.5 μm is indicated. This is probably a fundamental lower size limit for the basic units of life and if someone is going to look for life on Mars they should look for something larger than this.

This discussion has so far been focused on bacteria. Eukaryotic cells are always somewhat larger, mostly with sizes ranging from about 3 to 1000 μm. Eukaryotic cells contain a number of more space-requiring organelles (mitochondria, the cytoskeleton, the nucleus, centrioles, and often flagella) (see Chapter 8) that explain why they must necessarily be larger than bacteria.

We could also ask whether there is a maximum size for cells. It is practical to discuss bacteria and eukaryotes apart although in both cases the underlying reason is again one of diffusion. In bacteria, all transport of solutes to the cell as well as within the cell depends on molecular diffusion. Transport by diffusion is fast over very small distances and the time for transport increases with the square of the distance. Calculations show that metabolism and growth rates would therefore decrease substantially with increasing cell size, and bacterial cells do not generally exceed about 2 μm. A few kinds of giant bacteria (notably some sulphur bacteria and photosynthetic bacteria) usually contain a large interior vacuole. Eukaryotes have a cytoskeleton that allows for advective transport within the cell; also many eukaryotic microorganisms feed on particulate matter and are thus not dependent on diffusion for uptake of dissolved substrates. However, oxygen uptake depends on the diffusive flux from the environment. Simple calculations (based on metabolic rates, ambient O_2 concentration, and diffusion constants) show that an aerobic cell cannot exceed about 1 mm in diameter. When it comes to photosynthetic cells that depend on the flux of dissolved CO_2 from the environment, size is constrained to a diameter of 20–30 μm.

Multicellular organisms have evaded the fundamental size constraints through complicated internal advective transport systems, especially a circulatory system with a capillary network and an intestinal tract. As far as the individual cells making up animals or plants are concerned, the same size constraints apply as for unicellular eukaryotic organisms.

Life—from the viewpoint of thermodynamics

Classical thermodynamics predicts the direction of spontaneous processes. Such predictions often accord with our daily experiences and they are related to our sense of the direction of time. Heat, for example, flows from warm to cold regions and never the other way around. Another example is shown in Figure 4.1. A concentration gradient of dissolved molecules (e.g. a dye) will equilibrate over time due to molecular diffusion irrespective of the absence of stirring. There will be a net flux of matter from higher to lower concentrations. In any case this is only a statistical phenomenon. The individual molecules move in random directions at any time with a mean

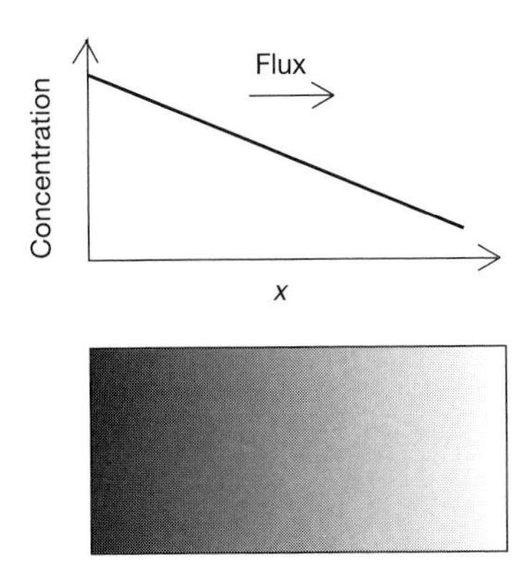

Fig. 4.1 Molecular diffusion. In a concentration gradient, net fluxes of a dissolved substance will be from higher to lower concentrations until the concentration is the same everywhere.

velocity that is a measure of the temperature. But if, at a given point in the container (Figure 4.1), the number of molecules moving towards left or right respectively are counted, then the latter will be more numerous, simply because concentrations (and thus numbers of molecules) are initially higher in the left side. If there were only very few molecules present, then one would always observe fluctuating concentrations everywhere. But in practice numbers of molecules are usually very high, so that diffusion can be described using exact laws, even though it is a statistical phenomenon.

These considerations exemplify the second law of thermodynamics. It states that a reversible process can only proceed spontaneously if the *entropy* of the system increases. Entropy can be given a more exact physical definition, but here it is sufficient to say that it is a measure of the probability of the state of a system. The direction of a process can thus be predicted. In the above mentioned example the situation where the molecules are regularly distributed has higher entropy than if all the molecules were found in one end of the container: the former situation is therefore more probable.

The second law is among the most successful results of physics. In our context, application to chemical reactions is especially important. The concept of *free energy* is somewhat related to entropy. A chemical process will run spontaneously only if the system has a negative change in free energy corresponding to an increase in entropy: it must be a 'downhill' process. A given process:

$$A + B \rightarrow C + D$$

will take place only if the free energy of the reactants (A, B) is higher than that of the products (C, D). Knowing the chemical composition of molecules, their concentrations, temperature (and in aqueous solution usually also pH) it is possible to calculate changes in free energy for all kinds of reactions. The energy that is liberated by a downhill process can be coupled to another process that carries a positive change in free energy as long as the net change of the overall process is negative.

Living organisms would appear to violate the second law of thermodynamics. An egg, for example, develops over time to become more and more complex and represent a very unlikely organization of matter and it appears to defy the universal tendency of dissolution of structure and complexity. At the beginning of the last century many biologists still felt that life represented an exception to the second law due to 'vital forces'. The analysis in Erwin Schrödinger's book *What is life* (1944) showed, however, that while it is (still) not possible to explain all vital processes by reference to physical laws, then it is possible to establish that life obeys all fundamental physical laws—including the second law of thermodynamics.

Classical thermodynamics is concerned with small changes in isolated systems close to their equilibrium. (By equilibrium is meant the composition of a chemical system when it reaches its lowest energy state; it is the state towards which the system finally approaches, and it is a function of initial concentrations of reactants, temperature, and pressure.) Life represents systems that are very far from equilibrium, and they are 'open' in the sense that they exchange energy and matter with their surroundings. The second law requires that the entropy of the entire system must increase, but it allows for a local entropy decrease as long as the entropy of the universe increases. An example is a frying pan filled with a sugar solution of suitable viscosity. If placed on the cooker the liquid will form small, often hexagonal convection cells. This represents a more improbable organization of the liquid. But the frying pan with the sugar solution is only an intermediate in an energy flow from the stove to the air in the kitchen. The entropy of the total system (stove, frying pan, air in the kitchen) will increase although this flux of heat accomplishes a local entropy decrease.

A couple of more relevant examples are shown in Figure 4.2. Above we consider a chemical process $A \leftrightarrow B$ with a temperature-dependent equilibrium value. The system is placed in a container that is heated at its left end and cooled at its right end. We further assume that the chemical equilibrium is shifted to the right at high temperatures and to the left at low temperatures. In this situation there will be a net transformation of A into B in the left side of the container and conversely, a net conversion of B into A in the right side of the container. It will also lead to concentration gradients and therefore fluxes of A and B towards left and right, respectively. The system will be maintained in a dynamic steady state at the expense of an energy flux through the system (adding heat on the left side and removing heat on the right side).

A system that is perhaps more analogous to life processes is shown in Figure 4.2. The left side of the box is assumed to be semi-permeable: it allows the small molecules X and Y to pass, but not the larger molecules A, B, and C. We further assume that the process $X \rightarrow Y$ is a downhill process that will only take place if catalysed by the enzyme A. Inside the box, X combines with A to produce B and Y. The metabolite Y diffuses out of the box and B decays through an intermediate product C back to A. The system obeys thermodynamic laws as long as $X \rightarrow Y$ implies a negative change in free energy.

Living organisms are systems that in some respects resemble this example. The maintenance of integrity and metabolism depends on the exploitation of free energy from the surroundings in the form of a chemical potential or light energy (that in

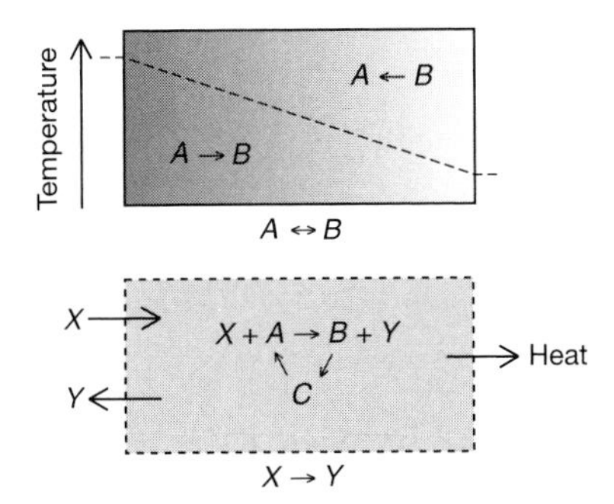

Fig. 4.2 (Top) A temperature-dependent process in a container with a temperature gradient (stippled line) will be maintained in a dynamic steady state because opposite processes will dominate in each of the sides of the container. There will also be a constant, oppositely directed net flow of the two reactants. (Bottom) The enzyme *A* is maintained behind a semi-permeable wall; it catalyses the transformation of *X* to *Y*. For further explanation, see text.

phototrophic organisms is quickly transformed into a chemical potential). Organisms are thus intermediates in an energy flux. When, for example, the food consumption of animals exceeds their growth this is because some of the energy thus obtained is needed for the maintenance of complicated structures and the integrity of the organisms; that is, to counteract the general tendency of dissolution and equilibration implied by the second law. Living organisms represent systems in a dynamic steady state. Schrödinger proposed the expression '[that organisms] feed on negative entropy'; this terminology has, however, not caught on. The metabolic processes of organisms will always lead to an entropy increase of the entire system (organism + surroundings) and thus a total decrease in free energy.

Living organisms maintain complex structures and cycling of materials due to an energy flux. They also maintain a chemical disequilibrium between themselves and their surroundings: organisms are generally chemically reducing relative to the environment. Green plants exploit electromagnetic radiation for the chemical reduction of CO_2 with H_2O to carbohydrates $[CH_2O]$ and with O_2 as a waste product. In this way they oxidize their environment, and being largely built from carbohydrates they become chemically reducing. This chemical potential that is given by the free energy change in the process $[CH_2O]+O_2 \rightarrow H_2O+CO_2$ is then exploited by a number of other types of organisms (e.g. by feeding on or degrading plant tissue). The thermal energy that results from all these processes will eventually leave Earth in the form of infrared radiation to space. We may therefore consider the entire biosphere as a link in the energy flow from the sun to the universe.

This thermodynamic description of life requires a note (that we have already implicitly assumed): the second law of thermodynamics can tell whether a process *may* occur spontaneously, but not whether it will actually do so or whether it will proceed extremely slowly (cf. the example in Figure 4.2: $X \rightarrow Y$ takes place only through the catalytic effect of *A*). It is very fortunate that not all thermodynamically possible reactions actually take place spontaneously. If this were so, a lump of sugar would spontaneously catch fire in our oxygen-containing atmosphere, and even worse—we would all end up as blue smoke in a second. The reason is that some processes need *activation energy* to proceed; processes may thus be *kinetically* constrained. In order to overcome such kinetic constraints so that processes run towards equilibrium it is possible to add energy (e.g. heat), but they can also be overcome through catalysis. In technical contexts so-called contact catalysts (such as amorphous platinum) adsorb molecules so that they react more easily at the higher concentrations. Enzymes are such catalysts in living cells. Enzymes are complex protein molecules and they typically catalyse only a single or a few related processes. The processes they catalyse are always thermodynamically possible, but through enzymatic catalysis cells can control which processes actually take place. Enzymes may also determine the particular pathway for a given net process. The oxidation of carbohydrates by an aerobic organism can be written as:

$$[CH_2O] + O_2 \rightarrow H_2O + CO_2.$$

The change in free energy can be calculated from this net equation irrespective of the actual pathway. In cells it takes place stepwise with many intermediate products and controlled by a large number of enzymes. Through these processes a large part of the released free energy is conserved as chemical energy in the form of adenosine triphosphate (ATP). This is an energy-rich molecule that is used almost universally in cells for energy-requiring processes (synthesis, motility, transport across the cell membrane). The energy of ATP is released by splitting off one (or two) phosphate molecules:

$$\text{ATP} \rightarrow \text{ADP} + \text{phosphate} + \text{energy}$$

(where ADP stands for adenosine diphosphate).

It should by now be clear that life with its complexity, chemical disequilibrium, integrity, and growth requires that it is an intermediate in an energy flux. Organisms depend on the access to free energy in their surroundings that ultimately is provided by electromagnetic radiation from the sun (and to a modest degree by radioactive decay in the crust of the Earth). Organisms can exploit this potential energy through the specific catalytic properties of enzymes.

It is also obvious that this description is incomplete and it is also somewhat ambiguous. First of all it does not explain reproduction or heredity. Also, there are non-living systems that also depend on energy dissipation. In addition to the theoretical examples in the beginning of this section it is obvious that other abiotic phenomena on the surface of the Earth can be quite complex: the hydrological cycle, climatic phenomena such as high and low pressures, fronts, storms, and thunder are also products of the energy flux through the atmosphere from the sun to the universe. Likewise geochemical processes are driven by geothermal energy in combination with atmospheric processes.

Autocatalytic cycles

The thermodynamic considerations can be somewhat expanded, and perhaps further approach a definition of life, by considering so-called *autocatalytic cycles*. A hypothetical example is shown in Figure 4.3. Here it is assumed that X combines with A to produce

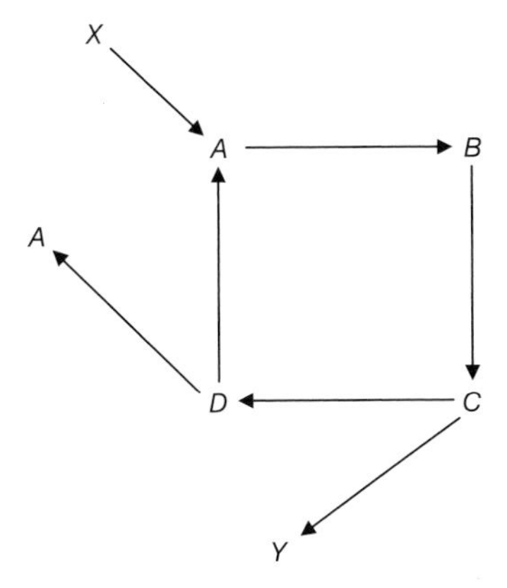

Fig. 4.3 An autocatalytic cycle. When X is added it will react with A to produce $2A$ plus the metabolite Y. For further explanation, see text.

B that is spontaneously transformed into C that decays into Y (a kind of metabolite) and D. Finally D is transformed into $2A$. The process is thermodynamically acceptable if $X \rightarrow Y + A$ represents a decrease in free energy. The difference compared with the models we considered in the previous section is that every cycle leads to a doubling of A. Therefore the new A can start its own cycle and the whole system will therefore grow exponentially as long as the combined energy and growth substrate X is in some abiotic way regenerated from Y or is generated in some other way in the environment. Two questions arise in this context.

The first question is whether such autocatalytic cycles are at all possible. The answer is that certain relatively simple chemical systems produced in the laboratory actually have similar properties. It is a problem that chemical processes are never completely 'clean'; various side-reactions will occur and result in by-products that are not functional in the cycle. The end-product of a cycle is therefore somewhat less than $2A$. But it is a necessary condition that the number of produced A is >1, otherwise the amount of A will decrease for each cycle until the process stops. The efficiency of the cycle equals the product of the efficiencies of each of the individual steps in the cycle. It is therefore difficult to make somewhat more complex cycles that will require specific catalysts to control the individual steps. This is perhaps a practical difficulty when attempting to understand the origin of life as a simple autocatalytic cycle, but it may not be a fundamental difficulty.

The other question is whether such cycles represent an adequate description of life. We certainly can say that autocatalytic cycles do represent a substantial component of a definition of life. The equation:

$$\text{bacterium} + \text{substrate} \rightarrow 2 \text{ bacteria} + \text{metabolites}$$

is identical to the model shown in Figure 4.3. But it is also not a complete description of life. The reason is that, as presented, the autocatalytic cycle does not include heredity and natural selection. We can imagine that A 'mutates' to a related chemical compound A^*. But even if A^* is still functional in the cycle there is no reason to assume that its 'offspring' will also become A^*. There is therefore no a priori reason to assume that simple autocatalytic cycles include heredity on which natural selection could work. So even if we have approached a somewhat realistic definition of life, something essential is missing. The definition of life necessarily implies two—one might say complimentary—aspects: a thermodynamic and a genetic one.

Life as replicators

A central property of life is that it is subject to natural selection—life is fundamentally associated with this mechanism. It is obvious that organisms as we know them could never arise spontaneously; only through Darwinian evolution is it possible to understand the development of the complexity of living organisms and this applies even to the simple bacterium. The Darwinian selection mechanism is a natural law (parallel to basic physical laws) that applies to replicators with the following three properties:

1 *Reproduction and heredity.* Units reproduce (e.g. by division); the properties of the individual units are inherited by the offspring.

2 *Variation and mutations.* A certain variation among the replicators is maintained. This variation results from heredity being not completely perfect (mutations: random changes in hereditary components).

3 *Differential fitness and selection.* Some hereditary varieties are more effective replicators than others (shorter generation time, lower mortality, etc.) and some varieties therefore increase in number relative to other varieties. Fitness is a function of the properties of the environment; therefore the result of evolution is that replicators (organisms) appear to be adapted to the environment in which they occur.

These three properties represent the basic essentials for evolution through natural selection. They must be present in order to characterize anything as 'life'. In addition to virus and to test-tube evolution of RNA molecules (Chapter 5) these properties are found exclusively in living organisms (although they can be simulated in computer models). Crystals have been used as an analogy to heredity (Figure 4.4). If a crystal is added to a saturated solution of the same compound the crystal will grow and perhaps, when the crystal has grown to a certain size, it may break up into smaller crystals that can continue growth. If the original crystal had some defect this is amplified during crystal growth and 'inherited' to its 'offspring'. As previously mentioned it has been suggested that the first genes ('crystal genes') were based on clay minerals and that the information encoded in crystal defects were later transferred to nucleic acids in an unspecified manner. However, such crystal genes show a very limited form of heredity and, in my view, the idea has only limited explanatory power when it comes to the origin of genetic systems.

The DNA molecule is the carrier of genetic information in all extant organisms. DNA molecules are built from nucleotides: phosphate–deoxyribose units with one of four organic bases (adenine, thymidine, guanine, or cytosine) as shown in Figures 3.3 and 4.5. The bases have a tendency to combine to a complementary base through

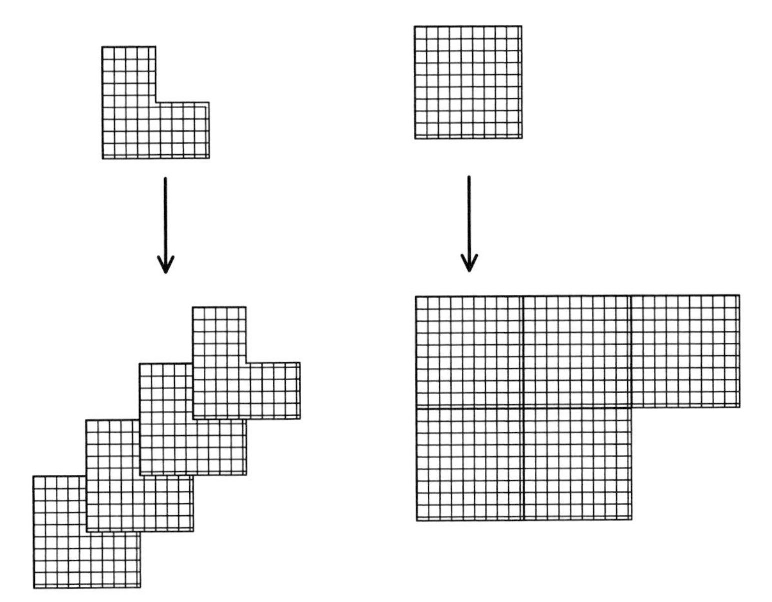

Fig. 4.4 Crystal genes. When a crystal with a defect (left) grows the defect will be 'inherited' by other crystals.

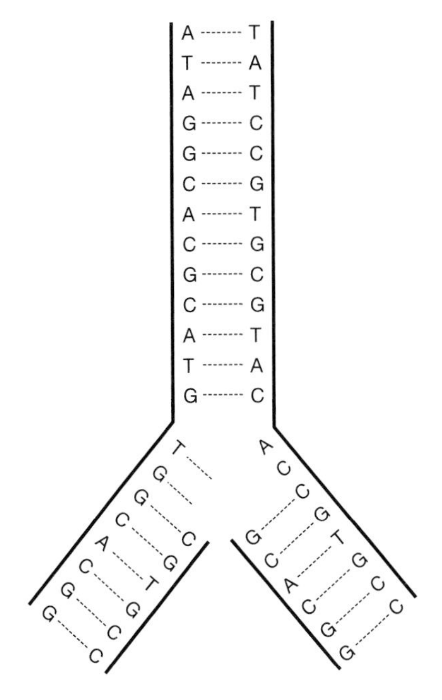

Fig. 4.5 Replication of double-stranded DNA.

hydrogen bonds so that A forms pairs with T, and C with G. Through enzymatic catalysis and the necessary building blocks, a single DNA strand will form a complementary strand and DNA is normally found as a double strand that is twisted together in a double helix. Certain enzymes can accomplish a doubling of the DNA molecule (the chromosome) typically in connection with cell division. In bacteria (with which we are so far mainly concerned) the single chromosome is circular and connected to the inside of the cell membrane.

The genetic information is given by the sequence of bases along the chromosome—rather like the information in a written text consists of the sequence of letters along the lines. A common metaphor is that the genetic code is a language based on four letters (A, T, C, G) that form words of three letters—referred to as *triplets*. Most 'words' specify an amino acid (several amino acids are specified by more than one triplet), but there are also codes that ensure that the molecule is read correctly—so-called *stop codes*. The translation from particular triplets to a given amino acid is referred to as the *universal code* because it is found in all organisms (in fact, deviations from the universal code have been found, but the term universal is still appropriate). A long sequence of triplets (hundreds to thousands) may code for an entire protein and such a sequence is referred to as a gene.

Mutations are changes in hereditary information. They are random and have no functional direction. It has been long known that ionizing radiation and certain chemical compounds increase mutation frequency and it is reasonable to assume that such factors (e.g. cosmic radiation and oxygen radicals formed within cells under natural conditions) are responsible for many mutations. Mutations do not represent an adaptive feature in organisms; in fact, most mutations reduce the carrier's fitness, or they are selectively neutral or almost so. All extant organisms have complicated mechanisms that serve to reduce mutations or their effects, such as 'proofreading' during DNA replication.

There are different types of mutations. *Point mutation* means that one base has been exchanged for another one, for example, T instead of A. *Deletion* means that a smaller or larger part of the chromosome is lost, and *inversions* and *translocations* mean that a segment of the chromosome has been inverted or moved to another part of the chromosome. *Duplication* means that an extra copy of a given gene is inserted in the chromosome. Such gene doublings have been extraordinarily important in evolution because the two initially identical gene copies may diverge over evolutionary time and eventually serve different, although often related, functions. This may have been the single most important mechanism by which genetic information has increased over evolutionary time (Chapter 6). By *horizontal gene transfer* is meant that genes are transferred from one individual to another, not by vertical inheritance, but between organisms that may be unrelated to each other.

In addition to the chromosome, bacteria carry *plasmids*, small circular DNA strands that occur in the cytoplasm. They may code for particular traits (e.g. resistance to antibiotics) and they are often subject to horizontal transfer between bacterial cells. Eukaryotic cells harbour two types of organelles (mitochondria and chloroplasts) each with their own bacteria-type chromosome. The reason for this is that these organelles originated as endosymbiotic bacteria.

Structure and function of a bacterial cell

At this point it may be useful to discuss how the simplest known organisms function. Most bacteria have shapes like small cylinders (rods), spheres (cocci), commas (vibrios), or corkscrews (spirilla), and others deviate from these categories. A typical bacterium measures about 1 μm (constraints on bacterial size range were previously discussed). Some types of bacteria are shown in Figure 4.6 (see also Plate 1) and Figure 4.7. They are typically covered by two cell membranes and a cell wall. The innermost (real) membrane corresponds to the cell membrane of eukaryotes. It is composed of a double layer of *phospholipids* into which different protein molecules are inserted (Figure 6.2); the proteins serve, among other things, as transport channels for dissolved molecules. Outside this membrane there is a rigid cell wall consisting of *peptidoglycans*. Outside of the cell wall there is a *periplasmic* space

Fig. 4.6 (See also Plate 1) (Top, left) Cultures of purple and green sulphur bacteria. (Top, right) The filamentous, colony-forming cyanobacterium *Pseudanabaena* and some other prokaryotic microorganisms. (Bottom) Microorganisms are not always invisible to the naked eye: the figure shows a shallow brackish bay (north of Copenhagen) totally covered by purple sulphur bacteria. (Originals)

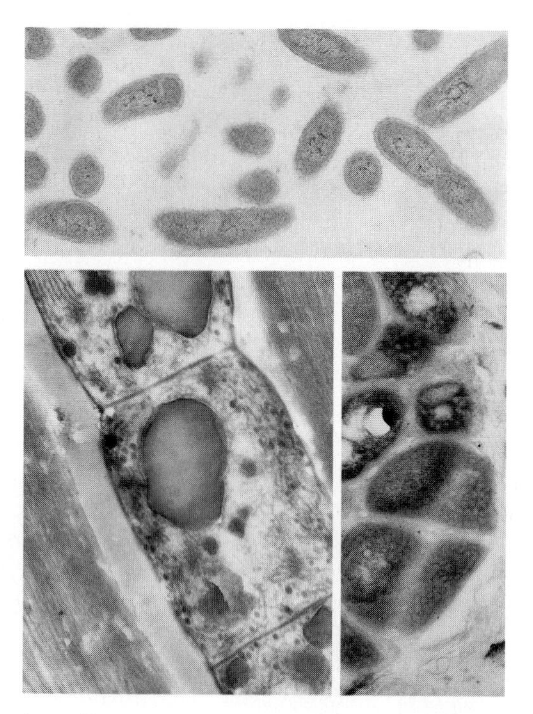

Fig. 4.7 (Top) Sections of bacteria in the electron microscope. The double membrane structure is visible and the black filaments are parts of the chromosome; a cell to the right has initiated cell division. Lengths of the cells are approximately 1 μm. (Bottom, left) A section through about 1½ cells of the cyanobacterium *Calothrix*; the diameter of the cells is about 2.5 μm. Photosynthetic membranes (thylakoids) are clearly seen in the upper left corner. The homogeneous grey areas are so-called carboxylosomes; they are stores of a photosynthetic enzyme. (Bottom, right) Parts of a colony of the purple sulphur bacterium *Thiocapsa*. The cells measure a little more than 0.5 μm across; the internal structure represents photosynthetic membranes. (Originals)

and finally an outer *lipopolysaccharide membrane*. There are several variations on this theme: the so-called *Gram-positive bacteria* deviate from this description and the so-called *archaebacteria* (a special group of prokaryotes that are not closely related to the *eubacteria*) deviate by using branched lipids (isoprenoids) in their cell membrane, and the cell wall also has a different composition. The cell wall is necessary because bacteria do not have a cytoskeleton like eukaryotic cells. The cytoplasm has a somewhat higher pressure than the surroundings and so the cell wall is necessary to hold the cell together. The absence of a cytoskeleton also means that bacteria cannot take up particulate matter or even macromolecules from the surroundings, but only low molecular weight compounds such as monosaccharides and amino acids or inorganic compounds. When bacteria utilize polymers as a substrate, these must first be degraded (hydrolysed) outside the cell by membrane-bound enzymes before the resulting monomers can be taken up through the cell membrane.

The inner membrane may have invaginations in some bacteria (especially in photosynthetic bacteria). Other structures that are visible in the microscope include liquid or gas-filled vacuoles. Finally *ribosomes* (see below) are found freely in the cytoplasm and attached to the inside of the inner cell membrane.

The total length of the bacterial chromosome is about 1 mm, but it is bundled very tightly to fit inside a cell that is only about 1/1000 mm. It contains 1–10 million base pairs, and codes for 1000 or more genes. There is usually only one chromosome, but some bacteria have two and others may duplicate the chromosome during rapid growth. The chromosome is circular and attached to the inner cell membrane at two sites. At one of these sites the chromosome starts its replication at a given stage of the cell cycle. When this happens, a new attachment site is formed on the new chromosome and it attaches on another site on the membrane. Due to the longitudinal growth of the cell the two new chromosomes depart from one another and they are finally distributed to each of the two daughter cells. The cell division is initiated by the formation of a septum in the middle of the cell before the two new cells separate.

Figure 4.8 shows schematically how the genetic apparatus functions within the cell. The genetic code in the form of individual genes or groups of neighbouring genes are *transcribed* into a single-stranded RNA molecule, so called *messenger RNA* (mRNA). This process is in principle like DNA replication (Figure 4.5) except that the base thymidine in DNA is replaced by uracil in RNA. The building blocks for RNA synthesis are nucleoside triphosphates (ATP, GTP, CTP, UTP). The additional phosphate

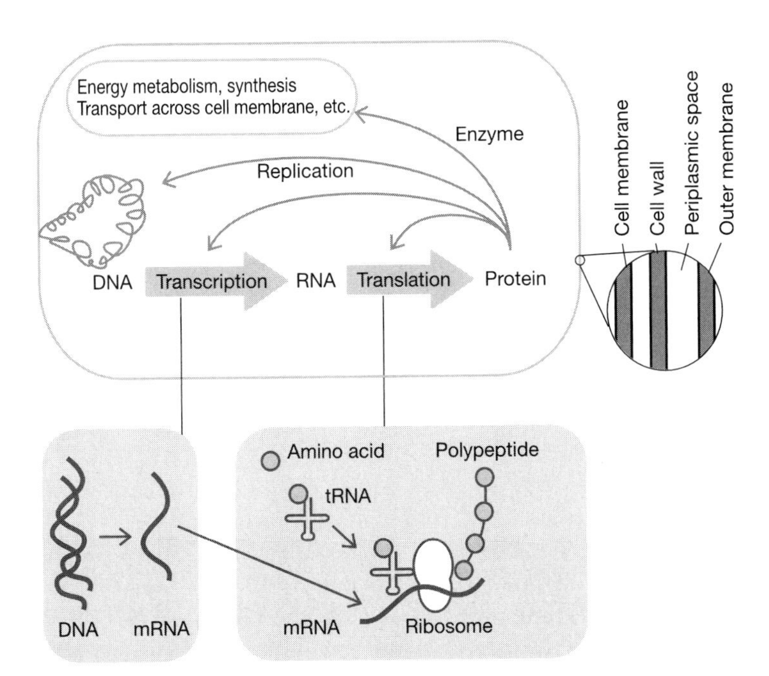

Fig. 4.8 The transcription and translation apparatus of a bacterium and a schematic presentation of the surface layer of the cell.

groups represent 'high energy bonds'; the splitting of these bonds by hydrolysis delivers the necessary energy for the polymerization of the RNA molecules and the process is catalysed by RNA polymerase.

Protein synthesis takes place in the ribosomes. They have a shape like a figure eight and consist of two types of RNA (rRNA) and protein. When the individual mRNA strands contact a ribosome the translation to protein can proceed. Proteins (polypeptides) are long, one-dimensional chains of amino acids. Amino acids are transported to the ribosomes with the aid of a third type of RNA called *transfer RNA* (tRNA). These are short (less than 100 nucleotides long) RNA molecules, partly with internal base pairing so that the molecules have a clover leaf shape. There are a little less than 100 kinds of tRNA in each cell and each is specialized to transport one particular type of amino acid that attaches to the 'stalk end' of the tRNA (Figure 4.9). At the opposite end of the tRNA molecule a sequence of three nucleotides represent an *anticodon* to the mRNA code for a given amino acid. The triplet GGU, for example, codes for the amino acid glycine and the anticodon is therefore CCA. Actually the genetic code is redundant, as a total of four triplets (GGU, GGC, GGA, and GGG) all code for glycine.

The anticodon triplet of transfer RNA molecules binds to the respective triplet on the mRNA strand and as it passes through the ribosome the amino acids are coupled together to a polypeptide strand and the tRNA molecules are eventually released again. The protein strand then folds up in a characteristic three-dimensional structure that depends on the amino acid sequence. Most proteins are enzymes and their specific catalytic properties are closely related to their three-dimensional shape.

It is obvious that the bacteria are poorly served if the entire genome is continuously transcribed and subsequently translated. Many enzymes are used only during certain parts of the cell cycle or in connection with particular environmental conditions, and when bacteria are starved, transcription almost comes to a halt. Some genes are therefore *repressor genes*: the proteins they code for prevent other specific genes (or groups of neighbouring genes, a so-called *operon*) being transcribed into mRNA. In this way the cell can regulate which genes are translated according to its physiological state or ambient conditions. There are also, of course, genes that specify the different types of tRNA and the constituents of the ribosome.

We may distinguish between two groups of proteins. Some catalyse processes that are related to the genetic system and to protein synthesis (e.g. regulation and catalysis of DNA replication, transcription to mRNA). The remaining proteins are responsible for all other functions of the cell that maintain its integrity and survival: proteins in the cell membrane that control import and export of solutes, enzymes involved in energy metabolism, enzymes that catalyse the synthesis of macromolecules, and structural proteins that constitute particular structures in the cell.

The duration of the entire cell cycle from cell division to cell division depends on the temperature and on the availability of substrate. It is typically about one hour under favourable conditions; under optimal conditions some bacteria have generation times that are shorter than 15 minutes.

Although simplified, this description of bacteria provides an impression of how complicated they are. We can consider the whole machinery as the way in which a

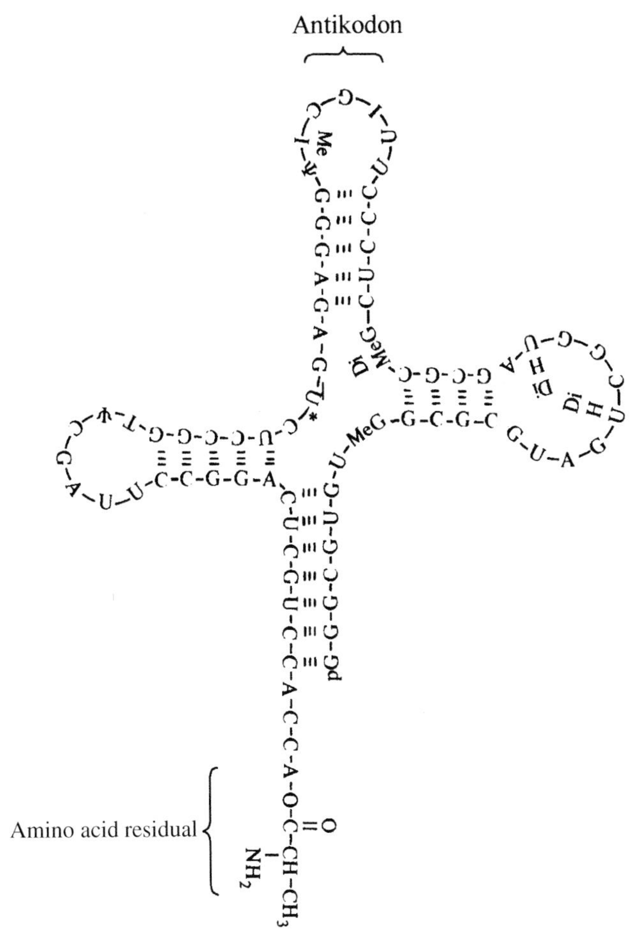

Fig. 4.9 Structure of tRNA for the amino acid alanine with the amino acid residue attached to it. I represents any base. The other symbols (besides A, C, G, and U) represent minor variations of the normal bases that otherwise occur in tRNA. The real molecule has a twisted three-dimensional structure. Redrawn from Miller and Orgel (1974).

DNA molecule produces more DNA. All the structures and functions that allow the bacterium to survive, compete with other bacteria, grow and divide is how the DNA molecule maximizes the probability of persisting into the future.

We can distinguish between two concepts that will be useful in the following narrative: the *genotype* and the *phenotype*. The genotype represents the genetic information that is encoded in the genome (the DNA molecule). It is the information that is hidden in the total sequence of A, T, C, and G. As is well known, it has recently become possible to sequence the genome of several microorganisms, and some plants and animals (including humans). But in all cases the function of the vast majority of the genes involved is still unknown. In principle, the information is that which would be required to construct an organism although this can only be realized in an intact cell. The genotype is in principle fixed, but rare errors (mutations) occur. The phenotype represents the appearance and properties of the organism in all respects. It is, first and foremost, a function of the genotype, but it is also influenced by the environmental conditions that a particular organism has experienced. Under all circumstances the phenotype can only be realized under some specified set of environmental conditions. Natural selection acts on the ability to survive, grow, and multiply in a given environment and this depends on the phenotype. A mutation that increases the chances of a phenotype surviving and multiplying will be favoured by selection, and its frequency in the population will increase over time.

We have seen that the genetic code is translated into a phenotype in the sequence:

$$\text{DNA} \rightarrow \text{RNA} \rightarrow \text{protein}$$

and never the other way around. This fact is referred to as the *central dogma*. While DNA may be influenced by external influences that lead to random mutations the reverse is not true, and neither the phenotype nor environmental impact on the phenotype can influence the genetic information in any directed way. Acquired attributes are not inherited! To be sure, the enzyme *inverse transcriptase* can catalyse the transcription of RNA to DNA; the enzyme is coded for in some types of RNA virus.

The previously mentioned 'chicken and egg problem' (Chapter 1) can now be rephrased as follows: Was there originally a phenotype that could accomplish the replication of a genome or was there first a genotype that specified a phenotype?

We have now seen a description (albeit simplified) of the function of the simplest type of organisms, as a background to some general considerations of the basic properties of life. The following two chapters deal with ideas about how life could have originated.

Chapter 5

Origin of life

The fact that no one knows how life originated is generally acknowledged, but the formidable difficulties in formulating even modestly reasonable models is perhaps appreciated less. And so many simplistic proposals have been made without addressing fundamental difficulties.

Understanding the physical–chemical conditions under which life arose, simple chemical processes in the environment that may have provided the energy for the earliest life, and the spontaneous ('self-organizing') mechanisms that might have created some sort of mechanical scaffolding (e.g. coacervates) for the earliest life are all important. It appears to me, however, that the most essential problem remains that of explaining the origin of replicators that were subject to Darwinian evolution. In this light the ideas about an RNA world will receive most attention here. While these ideas in no way solve all problems—on the contrary they tend to raise new ones—this approach is unique in that it is amenable to experimental analysis. This will be treated first. A second section will then be devoted to other largely speculative aspects of the origin of life.

The RNA world

A number of facts suggest that heredity was originally based on RNA rather than on DNA and that DNA's role in this respect is secondary. Due to the difficulty in synthesizing nucleosides in 'primordial soup experiments' it has been speculated that there was an even earlier, although chemically related, precursor to RNA that was a carrier of genetic information. This will not, however be considered further here.

Arguments in favour of RNA as the initial carrier of genetic information are the following:

1. RNA polymerization and replication can take place *in vitro* in a purely chemical system based on nucleotides.
2. The genome of certain viruses is based on RNA.
3. RNA is a more complex molecule than is DNA in the sense that single-stranded RNA can fold up in different three-dimensional configurations due to internal base pairing and so express a sort of phenotype (cf. the tRNA molecule). The double-stranded DNA molecule is a more stable molecule, but it can show very little phenotypic variation.
4. RNA can have a catalytic effect; so-called *ribozymes* catalyse cutting and splicing of other RNA molecules in extant organisms. The ribosome can also be considered as a ribozyme. Most recently, an RNA molecule has been produced

that catalyses the polymerization and replication of RNA. It can therefore be speculated that life arose as molecules that combined the geno- and phenotype.

5 The universal and central importance of RNA in all extant organisms, in particular in the context of protein synthesis.

6 Synthesis of deoxyribose—the carbohydrate backbone of DNA—takes place via ribose, the carbohydrate backbone of RNA. This can be interpreted to signify that DNA represents a further and later development of RNA.

7 Nucleosides function as coenzymes to many enzymes and as other biologically important molecules. An example is ATP (an adenine nucleoside with three phosphate molecules). This role of nucleosides can be interpreted as a relict from the RNA world.

All of these observations are interesting in their own right. But the most important aspect is that it is possible to study an RNA world in the laboratory. Such experiments initially took off in an RNA virus that attacks the bacterium *Escherichia coli*. The virus, referred to as Q_β, consists of about 4500 nucleotides. Around 1970 S. Spiegelman showed that RNA isolated from this virus will replicate in a test-tube in the presence of an RNA replicase isolated from the virus, magnesium ions, and the four nucleoside triphosphates ATP, UTP, CTP, and GTP (the phosphate bonds functioning as energy source). In order to follow the synthesis, one of the nucleoside triphosphates can be labelled with radioactive phosphorus; the RNA that is formed can be isolated, and the amount produced can be monitored from its radioactivity. Such an RNA culture could at intervals be transferred to a fresh medium so that its development could be followed over longer time periods (Figure 5.1). An interesting observation was that the RNA changed over time and lost its ability to infect bacteria; this was a demonstration of molecular evolution through Darwinian selection: the RNA had evolved to a new environment. The evolutionary change was, however, one of loss in genetic information. The need to synthesize the protein coat of the virus that is necessary for infecting bacteria was useless in the test-tube environment, and since shorter RNA molecules replicate faster, these were favoured by selection.

Further experiments carried out in Manfred Eigen's laboratory in Göttingen were initially based on an about 220 nucleotide long 'satellite RNA' from the Q_β virus. To some surprise it was shown that this primer was, in fact, unnecessary for the production of RNA from nucleoside triphosphates. If the system contained only the RNA replicase enzyme, the four nucleoside triphosphates, and Mg^{2+}, then roughly equally sized RNA strands appeared and replicated. This opened the opportunity to carry out selection experiments. Heating the systems gave rise to more heat-stable RNA molecules, and if ribonuclease (an enzyme that hydrolyses RNA) was added RNA appeared that folded up in a way that made it more resistant to hydrolysis.

The RNA molecules that were formed contained about a hundred nucleotides, and due to internal base pairing they have a hairpin-like shape. Eigen has coined the word *quasi-species* for these self-replicating molecules. A sort of RNA world had been created in the laboratory.

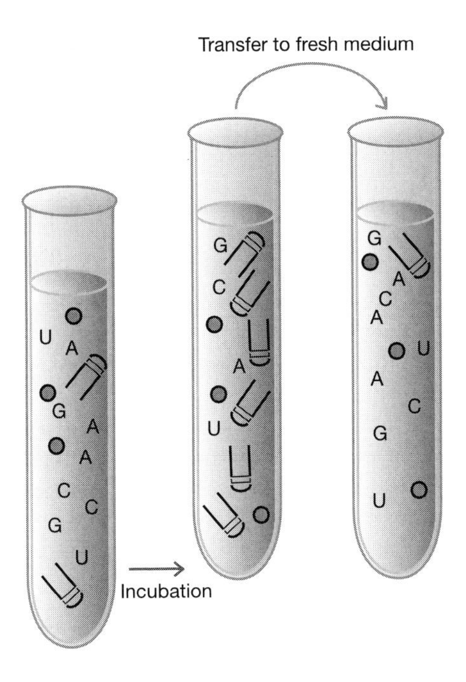

Fig. 5.1 Transfer 'cultures' of replicating RNA molecules. The medium consists of the four nucleotide triphosphates and the enzyme RNA replicase. When the medium is exhausted, part of the culture can be added to fresh medium for further growth.

But is this a complete model for the origin of life that we can now declare to be understood? Not really, and for several reasons. We will address these in turn in this and the following section. One difficulty is that reasonably confident replication of quasi-species depends on the presence of a complex protein enzyme, RNA replicase. The enzyme is necessary to maintain the 'life' of the molecules and it is unsafe to assume that such a complex enzyme would have formed 'accidentally' in a prebiotic world. The RNA molecules that did form are too short (they do not contain a sufficient amount of information) to code for such a large enzyme. We will return to the question of why only short RNA molecules develop in this type of experiment.

Of course the attempt has also been made to make RNA and to make it replicate in systems without catalysts or with simple, more non-specific catalysts. Leslie E. Orgel has made such attempts with replication of short RNA molecules in enzyme-free systems. Moderate success was attained with very short (less than 10 nucleotide long) RNA molecules, in particular in the presence of zinc or lead ions. RNA polymerase contains zinc; and one can speculate that this represents a relic from the primordial soup.

Recently, W. K. Johnston and colleagues have produced an RNA molecule that catalyses the replication of RNA, using selection experiments rather like those of Eigen. This is an important result, but even in this case the size of this artificially made ribozyme is much greater than that of RNA molecules (up to 14 nucleotides) that it can confidently replicate.

The sizes of the RNA molecules that form in such experiments are always relatively small and depend on the specificity or quality of the catalyst. And the successful catalysts seem to be much more complex than the replicators that are produced. This is a fundamental problem. It seems that selection actually favours smaller and less complex replicators. One reason may be that smaller molecules replicate faster. Another problem is mutations: the quasi-species are not very stable. In such RNA world systems there will always be some molecules that approach the maximum fitness, but also a large number of less fit molecules that are constantly formed through a high mutation rate. Mutations are of the same kind as those occurring in the DNA of real organisms: point mutations, deletions, etc. The probability of a mutation increases proportionately with the number of nucleotides per molecule. The mutational load of the population therefore increases with molecule size and mutation rate sets a limit to genome size. In the presence of the RNA replicase enzyme, quasi-species up to 100 nucleotides could form, and with a non-specific catalysis only very short RNA molecules could be generated.

A numerical example is useful at this point. Assume that there are three quasi-species in the primordial soup, and that they consist of three, four, and seven nucleotides, respectively. We further assume that the mutation rate is 0.1 per nucleotide per generation. For simplicity we also assume that mutated molecules have a fitness = 0; that is, all mutated molecules vanish before they can replicate again. In the first generation 27% of the three nucleotide quasi-species will be mutated (since $[1-(1-0.1)^3]=0.27$). For the four nucleotide quasi-species the corresponding number is 34%, and for the seven nucleotide quasi-species it is 52%. We see that the shortest quasi-species will win because its effective growth rate is highest. The largest (seven nucleotide) would become extinct irrespective of competition with the shorter ones, because it leaves on average less than one progeny per generation. We see that mutational load limits the genome size and thus the possibility for the evolution of a sufficiently large molecule that could accomplish a confident replication of larger RNA molecules.

Eigen has attempted to solve this dilemma by suggesting a model (Figure 5.2) he has called a *hypercycle*. It is a sort of autocatalytic cycle in which the replication of, for example, three quasi-species *A*, *B*, and *C* are coupled so that *A* catalyses the replication of *B*, *B* catalyses the replication of *C*, and finally *C* catalyses the replication of *A*. This would stabilize the system and force all three components into the same effective replication rate. If, for example, there is an excess of *A*, this will increase the replication rate of *B*, and if *C* is rare this will slow the replication of *A*. More genetic information could thus be maintained because the smallest replicators would not automatically be favoured by selection.

There are problems with this model. First of all, the above mentioned system may be stable in the short-run (in an ecological time-scale), but it will be evolutionarily

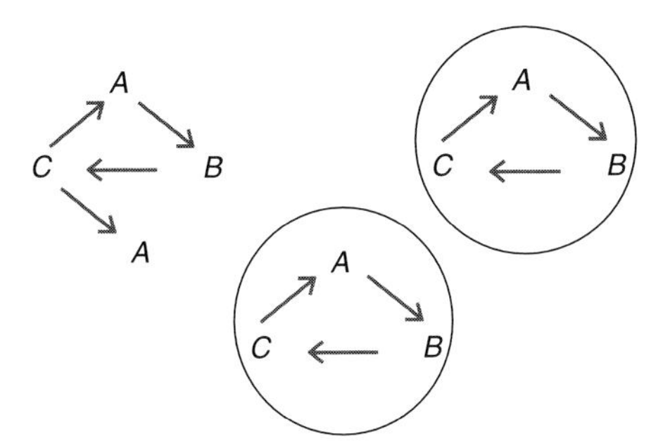

Fig. 5.2 Models of hypercycles. (Left) Mutations in *A* that improve the catalytic effect on the replication of *B* will not be favoured by natural selection because this will benefit all *A*'s in the solution. (Right) If the cycles are compartmentalized, then the cycles will have become units of selection and select for 'altruistic' behaviour of *A* because this will not benefit competing *A*'s.

unstable. This is because it requires an 'altruistic' behaviour of all components. If we consider only one of the quasi-species, we can imagine three types of mutations that will affect the hypercycle. Thus *A* could mutate so that its own replication rate increases, it could mutate so that it would more effectively exploit the catalytic properties of *C*, and finally it could mutate so as to improve the replication rate of *B*. The two former types of mutations would increase the fitness of *A* and would be adopted. The third type of mutation, affecting the mechanism that holds the cycle together will not, however, be automatically adopted. This is because such a mutation will not only favour the particular carrier of the mutation, but also all other *A*'s in the population. Selection acts on individual fitness relative to conspecifics and not on 'good for the species'. If such a mutation would imply any price for *A* by decreasing its replication rate, then selection would favour egotistic behaviour—which means favouring its own replication rate and its exploitation of *C*, but not contributing to the stability of the cycle. Natural selection will thus act to destabilize the hypercycle.

The way out of this problem is to assume that the individual cycles somehow became separated spatially, that is, they become compartmentalized (Figure 5.2). This could happen if they were enclosed in a membrane bubble of some sort (such as a double lipid membrane, a precursor of a cell membrane, or some sort of coacervate). In such a case, a mutation in *A* that increases *B*'s replication rate may be favoured because this will (via *C*) eventually favour the particular specimen of *A* and not all the other *A*'s in the population. Such a compartmentalization would mean that the entire compartment ('cell') would become a unit for natural selection. Irrespective of the degree of realism in this sort of model, it is clear that compartmentalization of such primitive genetic systems must have taken place very early as a prerequisite for the evolution of further complexity.

There are other difficulties with the hypercycle model. The components of compartmentalized hypercycles could still evolve into parasites or 'cheaters'. It is easy, for example, to imagine that a short circuit of the hypercycle (e.g. a mutation in *B* that directly catalysed the replication of A) could be favoured by selection. Also one of the components could become a parasite even in compartmentalized cycles, using only the catalytic capabilities of other quasi-species, but not catalysing their replication. This could be favoured selectively because the parasite could then decrease the size of its genome and thus increase replication rate. This would, of course, result in a collapse of the cycles. It is necessary to assume that the environment in some way favoured increasing genomic sizes and complexity of such hypercycles; otherwise it is difficult to see how they avoided degeneration into the smallest possible replicators.

We have so far discussed hypercycles in a rather abstract manner. But they do have some attractive similarities to what goes on in real cells. Figure 5.3 shows a somewhat more concrete model. It consists of two RNA quasi-species 1 and 2. Each consists of a + and a – strand. We could imagine that the – strand of 1 acts as a catalyst for 2 and vice versa. Such a system could be a kind of precursor for a (still protein-free) system with an mRNA (the + strands) and a tRNA (the – strand).

We have seen that the mutational load sets a limit to the size of RNA quasi-species. This is a general property of replicators including real organisms. There will always be an upper limit of mutational load (and thus genome size) above which populations of replicators crash. The higher the number of nucleotides the higher is the probability of dysgenic or lethal mutations. It is possible to formulate exact equations for the upper limit of genome size as a function of mutation rates and of the decrease in fitness of mutants relative to the optimal genotype. The mutation rates of extant organisms are much lower than those of more or less specifically catalysed experimental RNA replicators. In real organisms low mutation rates are due to complicated enzymatic machinery. But genome size is in principal always limited by mutation rates. This in part explains the genome sizes of different organisms and viruses. This can be seen comparing numbers of base pairs with mutation rates. The DNA molecule is more stable than the RNA molecule, and the double-stranded nature of DNA also allows for a proofreading during replication. Comparing RNA and DNA viruses, the former have a mutation rate per generation that is about 5×10^{-4} per nucleotide and genome lengths of about 5×10^{3} nucleotides. In the case of DNA viruses the corresponding numbers are about 10^{-8} and 5×10^{4}. In bacteria the mutation rate

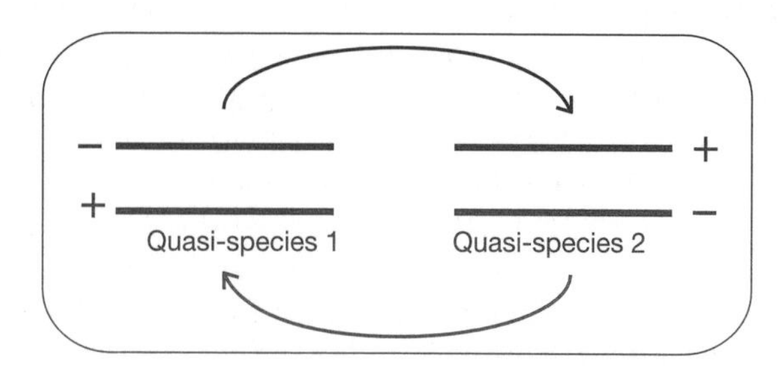

Fig. 5.3 A simple theoretical hypercycle involving two RNA quasi-species. Here the – strand represents a sort of tRNA that catalyses the + strand of another quasi-species, and vice versa.

per base pair has been estimated to be about 5×10^{-10} and the genome length to about 5×10^{6} base pairs. In eukaryotic cells the genome is 10–1000 times larger although mutational rates are not different from those of bacteria. However, in eukaryotes a large part of the chromosomal DNA is not translated (Chapter 8) and therefore insignificant. The translated part of the genome of eukaryotes is generally between 5–50 times larger than in bacteria. In addition to mutational rates there seems to be one other mechanism that limits bacterial genome lengths. In prokaryotes the chromosome can replicate only at one site at any time and so the time taken to replicate the chromosome is limiting. In contrast, the chromosome of eukaryotes can divide simultaneously at several sites. But under all circumstances, the mutation rate ultimately limits the size of genomes and thus the complexity of organisms.

Origin of life—chance or necessity

The RNA world hypothesis presents a real advance in attempts to understand the origin of life. Its primary strength is that it has provided an experimental tool and this is essential for further progress. It also solves the dilemma of whether a genotype or a phenotype was first. In the RNA world, quasi-species are simultaneously geno- and phenotypes and inheritance is direct. There are also many difficulties inherent in the RNA world. For example, it is unknown how nucleosides could have been synthesized in a prebiotic world given current ideas on the chemical environment on prebiotic Earth; at least it has not, so far, been possible to synthesize nucleosides in any primordial soup-type experiments. As the discussion in the previous section shows, it is also not trivial how more complex hypercycles could have become stabilized.

There are two schools of thought when it comes to the origin of life: 'genetics first' and 'metabolism first'. The 'genetics first' school is represented by Eigen and by Maynard Smith and Szathamáry, and the 'metabolism first' school by Oparin, Wächterhäuser, Cairns Smith, and Dyson (see Chapter 3). To a large extent this dichotomy is artificial and perhaps mainly a semantic problem, that is, a question on how life is defined. It is obvious that the putative RNA world could not have started in a vacuum. In addition to the abiological synthesis of nucleosides it would primarily have required some sort of environmental energy transduction that generated high energy phosphate bonds as polyphosphates or nucleoside triphosphates from electromagnetic radiation or from some sort of chemical reaction. So-called self-organizing mechanisms leading to phenomena such as coacervates, or lipid spheres that could be the precursor for cell membranes would also be a necessary pre-condition for the spatial ordering or compartmentalization of the first replicators. Perhaps energy transduction of some sort became a property of these membrane precursors before replicators arose. The 'genetics first' view simply holds the view that for anything to be defined as 'life' it must have replicator properties and be subject to Darwinian selection. Without these properties similarities with real life are superficial. Many of the ideas and experimental results on possible pre-genetic 'evolution' are ingenious and interesting and may contribute to understanding how replicators arose and became established. But without replicators they do not explain the origin of life and they have no explanatory power regarding further evolution.

Was the origin of life a fortuitous event or will life always develop under a given set of chemical and physical factors that prevailed on early Earth? The question cannot be answered. It can be stated with certainty that life's later evolution and the life we know is the result of an endless number of contingencies. Evolution includes both chance and causal mechanisms as expressed in the title of Jacques Monod's book (1970): *Le hasard et la nécessité*. The 'necessity' is due to the constraints given by already established structures and functions (such as, for example, the genetic code and the protein synthesis machinery that is common to all organisms) and by Darwinian selection. By 'chance' is meant the directionless mutations and a variety of other contingencies (stochastic processes, climate changes, etc.).

To mention a couple of examples: Australia was already isolated from other continents when placental mammals originated. The mammals of Australia are therefore all marsupials that underwent an adaptive radiation much as the placental mammals did in the Old World. We can recognize different types of marsupials that fill similar ecological niches as do placental mammals elsewhere. Names such as the Tasmanian wolf, marsupial anteater, etc., describe this. These animals often appear surprisingly similar to their Old World counterparts although they are the result of independent evolution. But the precise convergence does not always apply: kangaroos represent the ecological counterpart to ruminants (deer, buffalos, etc.) in other continents. But here the external similarity is limited: large herbivorous mammals with a ruminant-type digestion do not necessarily come out as something resembling a cow. Another example is the extinction of the dinosaurs. They all (except for birds) disappeared at the end of the Cretaceous period—either due to the aftermath of a collision between Earth and an asteroid or because of global cooling during this period. Both events are fairly well established and a number of other animal groups also became extinct in the period; but the identity of the decisive event is still disputed. At any rate, thanks to the extinction of the dinosaurs mammals could diversify and take over a number of ecological niches. And had the dinosaurs not gone extinct then we would probably not be around today.

Evolution is thus full of contingencies. If a large number of planets identical to he prebiotic Earth had all been seeded by one kind of bacterium then, after a few billion years, evolution would have taken quite different paths on the different planets. The fact that there are humans on Earth today is a phenomenon that four billion years ago (and even much later) would have had a vanishingly small *a priori probability*—our existence depended on the right number coming up in Monte Carlo. A priori probability can be explained in the following way. Suppose that a large city includes one million registered cars. The probability of seeing a particular licence plate on the first car one meets going out on the street is, of course, low (one in a million). Conversely, if someone goes down on the street without any beforehand expectations and the first observed car happens to have licence plate number PK34875, then this does not really appear as a miracle—although seeing this particular licence plate number had a very low a priori probability.

The earliest evolution of life must also have been characterized by a number of contingencies that established basic structures and functions such as, for example, the genetic code (see Chapter 6). The fact that all extant organisms have a common origin (Chapter 13) could be interpreted to mean that life arose only once (and therefore represented a very improbable event). It is possible that only very limited possibilities with respect to structure and function can form the basis for life. Another possibility is that life arose several times independently (perhaps including different fundamental functional types) but that all but one type became extinct through competition or for other reasons. Nothing of this can be answered today. It is also not possible to say whether the origin of self-replicating molecules (RNA or something else) in some sort of primordial soup, on mineral surfaces or whatever has a very low a priori probability. Even if such an event is probable in a geological timescale, and for example, could be expected to occur say once every 100 000 years in some specified system, this cannot be expected to occur in a laboratory experiment carried out by someone with a three year scholarship. There may be fundamental limitations that mean that we will never obtain a full understanding of how life arose.

Chapter 6

From the RNA world to the first cell

What are the problems?

Evolution from the (hypothetical) RNA world to the first bacterial cell is in many ways as difficult to explain as the origin of life itself. It is possible to formulate problems, but their solution seems in many cases to be almost insurmountable. Comparative studies combined with palaeontology can provide convincing descriptions on how, for example, Devonian lobe-finned fish developed into terrestrial vertebrates. But going from simple ideas on how life could have originated and relatively simple chemical systems with replicating RNA molecules to the first cell represents a huge gap with little but speculation to fill it. A bacterial cell represents an immensely complicated structure in comparison with our ideas about an RNA world. All extant free-living bacteria—irrespective of their ecology—have at least about 1000 genes and this was the case relatively soon after life originated, maybe four billion years ago. Selection will always tend to favour loss of redundant parts of the genome because this will speed up replication and reduce mutational load. We are therefore forced to assume that it takes the level of complexity found in extant bacteria to make cells that are viable in different habitats on the surface of Earth.

Intracellular parasitic bacteria like *Rickettsia* (typhus is a disease caused by a *Rickettsia* species) have smaller genomes than free-living bacteria because they can exploit many functions of the host cell, which also provides a relatively stable and predictable habitat. The ultimate evolution of an intracellular parasite is to become a virus. Viruses have only a couple of genes and otherwise depend entirely on the metabolic machinery of the host cell. (Viruses probably originated independently during evolution from intracellular parasites, but also from detached genes of various types of organisms including the viral hosts themselves.) RNA viruses in some respects resemble and share properties with the experimental RNA world and with ideas of a Precambrian RNA world. Perhaps RNA viruses even represent descendants from the RNA world. But we do not understand how the necessary resources and environmental requirements for the hypothetical RNA quasi-species could have been provided in the absence of living host cells with their enzyme machinery and energy metabolism. While we can imagine how bacteria can degenerate into a sort of parasitic RNA quasi-species, it appears almost impossible to understand how the RNA world could ever have developed into a cell.

The evolution of metabolic processes will be discussed in the following chapter. Here we will consider the following problems:

1 How proteins took over the dominating role as enzymes from ribozymes, how different types of RNA were differentiated, and the formation of a chromosome.

2 Explaining the origin of the genetic code.
3 Explaining why natural proteins contain only 20 kinds of amino acids (among a much larger number of possibilities) and why only L-amino acids occur.
4 The shift from RNA to DNA as carrier of genetic information.
5 The relationship between genetic information and the phenotype and the increase in genetic information.
6 The origin of the cell membrane.

Proteins as enzymes RNA quasi-species are self-replicating molecules that are simultaneously geno- and phenotypes. In the RNA world inheritance is direct. Introduction of proteins as enzymes resulted in a much larger enzymatic repertoire and a more specific catalysis. At the same time the proteins represented the first phenotype that is clearly distinct from the genotype. Inheritance of the phenotype thus became indirect.

Nothing is known about how proteins entered the scene. But it is permissible to speculate: the enzymatic capability of the RNA strands could be improved if individual amino acids were attached as in tRNA of real organisms. That is, amino acids, and perhaps later groups of amino acids acted as coenzymes for the ribozymes. The next step could have been a specialization of RNA so that the + strand had the role as mRNA and the – strand functioned as a tRNA that attached to the + strand with an anticodon triplet. Finally, the amino acids could be coupled together as a polypeptide strand that would further improve catalytic activity.

Two observations may, perhaps, give some support to these ideas. We may consider a tRNA as a sort of 'handle' on amino acids. In extant organisms many enzymes also have a 'nucleoside handle' and this can perhaps be considered as a relic from such an evolution. The other observation is based on nucleotide sequencing of different tRNAs that allow for the construction of a phylogenetic tree (Chapter 13). Eigen has emphasized that tRNAs with similar function (i.e. binds to the same amino acid) are more closely related (irrespective of whether they come from a bacterium, a yeast, or a fruit fly) than are different types of tRNA from the same organism. This suggests that tRNA can be traced back not only to the last common ancestor of all living things, but also back to the origin of the genetic code and to an RNA world. The origin of mRNA, and later the RNA–protein complex constituting ribosomes, could be imagined as a further specialization of different RNA molecules.

It is not difficult to imagine that functional specialization between different RNAs was adaptive in increasing the efficiency of the hypercycles within some sort of protocells. Some kinds of RNA specialized in collecting amino acids and others in coupling them together on the basis of the code in a third kind of RNA. A tendency for specialization applies to all levels of biological organization. Eukaryotic cells are internally compartmentalized and the different parts have different functions. At the multicellular level cells are differentiated to perform different tasks. At the ecological level the fundamental reason why there are so many species is also that specialists to some degree are more efficient than generalists. The observation that 'A jack of all trades is a master of none' may also have applied in the RNA world.

It is also conceivable that once quasi-species had been enclosed in some sort of membrane bubble and formed hypercycles it would become adaptive to have the genes linked together in a single chromosome. The RNA molecules replicated inside their protocells, but the protocells themselves must have divided periodically. As long as there were several different kinds of free-floating RNA molecules, a cell division must have represented a sort of lottery in terms of which molecules ended up in one or the other daughter cell. The relative proportion between the different kinds would then, perhaps, not be optimal and a single vital type of molecule might have been absent leading to daughter cells with reduced fitness. Connecting the RNA molecules into a single strand combined with simultaneous replication would then secure that each cell received the equal distribution of the genome between daughter cells. Also, linking genes together in a chromosome would limit inter-gene competition so that the entire chromosome became a more stable unit for natural selection.

The origin of the code As mentioned earlier, the code is based on four letters: C, G, A, U and three letter words. There are therefore 64 different combinations. Among them, three are stop codes (UAA, UAG, UGA). Since there is only need for specifying 20 amino acids the code shows redundancies. Thus the most common amino acids are each coded by four different triplets (e.g. glycine GGX, and alanine GCX, where X stands for any of the four bases). Generally the first two bases in a triplet seem most important whereas the last base in many cases is without importance. The code is almost universal. There are some small deviations where a code specifies another amino acid; this applies to animal mitochondria and to ciliates. How this happened is disputable, but it is certainly a secondary phenomenon.

How did the code originate? This question cannot yet be answered. There is apparently no relationship between the code and the chemical structure of the amino acids. In that sense the code is arbitrary: there is no chemical reason why a given tRNA binds to a particular amino acid in one end of the molecule and has a given triplet code in the other end of the molecule. One could ask why there are four bases, why the code is expressed in triplets, why there are redundant codes, and finally why only 20 amino acids are used. The best answer that can be offered today is that all these features minimizes the risk of replication errors. Using more than four bases would lead to more mutations. A doublet code would be insufficient to code for all amino acids whereas longer words (four or more bases) again would increase the risk of replication errors. The fact that the code is redundant also decreases the effect of point mutations because the substitution of a base in many cases would be ineffectual. A larger number of amino acids would also increase the error frequency, but why it is exactly 20 cannot be answered; fewer would have allowed for a doublet code.

The reason why all organisms use only L-amino acids is a classical question. Except to note that L- and D-amino acids may have different chemical and physiological effects, there is no reasonable explanation at hand.

To all these questions there is—besides of selection to minimize mutation rates or the effect of mutations—only the general answer that the code and the preference for L-amino acids represented solutions that, perhaps entirely accidentally, were fixed at an early stage of the RNA world. Once these traits had become established, it was impossible to change them because any deviation would become fatal. Technological devices may serve as an analogy. Their future development is often constrained by the original design. However, they might have been constructed in a quite different (and perhaps better) way if someone started building a device with a similar purpose all over again from scratch.

Real answers to these questions may perhaps only be available if we ever have the opportunity to observe life that has developed independently of that on Earth or if it will be possible to induce an experimental RNA world to produce proteins. Neither is very likely to happen in the foreseeable future.

From RNA to DNA It is easy to understand that the emergence of complex organisms was conditioned on the transfer of genetic information from RNA to DNA. DNA is much more stable and the replication of DNA is much more complicated than that of RNA. DNA predominantly occurs as a double strand and this allows for a kind of enzymatic proofreading and correction in connection with replication. The mutation rate associated with DNA replication is therefore much lower than that of RNA. For reasons discussed in the previous section an RNA organism could only contain genetic information corresponding to 1000–10 000 base pairs before reaching an error threshold, but a bacterial chromosome already carries 1–10 million base pairs.

It is much more difficult to understand exactly how the transition from RNA to DNA as carrier of genetic information took place. The difference between RNA and DNA is that uracil is replaced by thymidine; also the carbohydrate backbone of DNA is based on deoxyribose rather than on ribose. In cells, deoxyribose is formed through an enzymatically controlled reduction of ribose. This is taken as evidence that DNA came after RNA. Exchanging DNA for RNA must also have required the enzyme inverse transcriptase, an enzyme that transcribes RNA into DNA. Today we know this enzyme from RNA viruses that are transcribed to become a part of the chromosomal DNA of host cells.

Increasing genetic information during evolution It is commonly believed that evolution represents a continuous increase in complexity. That is not the case; selection often leads to loss of genetic information and secondary simplification. There is also evidence to suggest that genome sizes within different major organizational levels have been relatively stable over long geological periods and that major increases in genome size seem to have taken place in a stepwise manner (in a geological time-scale). We have seen that mutation rates and (in prokaryotes, at least) time used for replication, limits genome sizes. Bacteria have 1000 or more translated genes; for unicellular eukaryotes, simple invertebrates (e.g. fruit flies), and higher vertebrates (including humans) corresponding numbers are a couple of thousand, about 5000, and about 30 000 translated genes, respectively. Vascular plants are comparable with vertebrates in this respect. To those whose ego is disturbed by the thought that we are only 30 times more complex than bacteria and only six times more complex than fruit flies it may be a consolation that complexity (although not easily measured) is not linearly proportional to the number of transcribed genes. The produced proteins interact in several processes, with each other and with the expression of other genes in a complicated network, and so complexity grows rapidly with a modest increase in the number of translated genes. Under all circumstances, there is no reason to assume that the genome sizes of bacteria have increased much since the first real bacterial cell arose. Even the evolution of eukaryotes, and later multicellular organisms, was not associated with a drastic increase in the number of translated genes. Such a drastic increase must, however, have taken place during the earliest evolution of the first real cells.

The most important single mechanism for increasing genetic information is probably gene doubling. In itself this does not represent a qualitative innovation. But it allows for one of the gene copies, through mutations and selection, to take over new and more specialized functions. Such gene doublings have thus lead to the production of new enzymes and other biomolecules although they are usually somehow related to the functions of the original gene. In this way whole families of related genes co-occur in organisms. The family of haemoglobin genes in vertebrates provides an example. The original haemoglobin gene has doubled four times during the evolution of vertebrates resulting in myoglobin (that functions as an oxygen reservoir in red muscles) and in four types of haemoglobin molecules that all occur in, for example, humans. A large number of such gene doublings followed by functional specialization are known and so there are large families of structurally, functionally, and genealogically related biomolecules. Some of these gene doublings can be traced back to the common (universal) ancestor of all extant organisms (Chapter 13).

Bacteria may increase (or change) their genetic information in other ways. DNA can be transferred from one bacterium to another through horizontal gene transfer. This can happen in several ways. Bacteria can take up single DNA molecules from their surroundings (*transformation*); these DNA strands derive from other bacteria that have undergone lysis. Genetic material can also be inserted in cells via virus (*transduction*) or through direct transfer between two bacteria in physical contact (*conjugation*). This latter process is the closest bacteria come to sexual processes as known in eukaryotes. It may lead to *recombination*, that is, exchange of homologous parts of the chromosome between two cells. This process occurs with varying frequency in different bacterial species, or it may be entirely absent, and is probably a relatively rare event. Altogether large parts of the chromosomal genome of bacteria appear to have been very stable over evolutionary time, in particular with respect to the fundamental genes such as those involved in protein synthesis.

Prokaryotes also harbour plasmids. These are typically circular and short (1000–10 000 base pairs) pieces of DNA. They occur freely in the cytoplasm and are not a part of the chromosome. The delimitation between virus and plasmids is not sharp. Plasmids are not vital to bacteria, but often code for phenotypic traits that are adaptive under particular conditions. Resistance to antibiotics, production of toxins, and the ability to fix atmospheric N_2 are properties associated with plasmids. Plasmids are easily transferred from one cell to another by the mechanisms described above. Donor and receptor bacteria may be quite unrelated.

Endocellular bacterial symbionts have made a profound evolutionary impact, in particular with respect to the evolution of eukaryotic cells. Mitochondria (responsible for the energy metabolism in almost all eukaryotes) and chloroplasts both have their origin as endosymbiotic prokaryotes. There is also, if somewhat less compelling, evidence that other types of chimera formation played a role in the early evolution of the eukaryotic cell (see also Chapter 8). Such events lead to an increase in genome size and in complexity. During the subsequent evolution, many of the symbiont genes ended up on the chromosomes of the host cells through horizontal gene transfer.

The nature of genetic information

The significance of the genetic information contained in DNA is often misunderstood. In part this stems from spates of reports claiming to have found genes for various diseases and even for alcoholism, aggressive behaviour, and a variety of other things. Quite besides of the fact that these sorts of investigations may be wanting in rigour and that their presentation in the media is even worse, they also give rise to misunderstanding of the nature of genetic information. There are, of course, *alleles*—or more frequently—a combination of alleles that lead to or at least correlate with certain phenotypic traits such as the risk of getting certain diseases, body size, eye colour, and so forth. (By alleles is meant different versions of genes that are situated in a particular site on a chromosome.) But in reality there are no genes labelled 'eye colour', 'becoming a dwarf', or 'left hind leg'. Genes only specify proteins (enzymes). Different variations of a gene (alleles) can, through variations in the specified protein, have larger or smaller effects on the phenotype. The majority of 'disease genes' are alleles that produce malfunctional proteins or are not translated at all. But chromosomes do not represent a blueprint for organisms in the sense that there is a one-to-one correspondence between individual genes and different parts or functions—as is the case for an architect's drawing of a house.

The genetic code is arbitrary in the sense that there is no physical–chemical reason why given triplets code for given amino acids. The genetic code is also a symbolic language somewhat analogous to human languages. A word like 'squirrel' gives the association of a rodent with red fur and a bushy tail. But we might as well have used the term 'bicycle' to refer to this mammal. Likewise a computer file is a totally meaningless sequence of 0's and 1's that cannot be meaningful unless one has a program that can interpret it. And this program is again based on arbitrary conventions and cannot be reconstructed on the basis of a file written for it. Similarly, DNA represents symbolic information, and without knowing the 'language' it is just a meaningless sequence of A's, T's, C's, and G's and DNA is entirely ineffectual outside a living organism. Maynard Smith has asked the question whether such a symbolic, and in principle unlimited system for containing information and for heredity, is a necessary condition for anything we can call life, but the question cannot be answered yet.

There are cases where the relation between a certain allele and its phenotypic effect is known. This is typically the case when the produced enzyme is defunct or is not translated and a phenotypic consequence (or at least one of several consequences) can be explained. But in general we are infinitely far from being able to explain how the translation of thousands of genes leads to a giraffe, a bacterium, or a toadstool. The production of a given phenotype is certainly in accordance with physical and chemical laws, but the entire network of interactions between genes and enzymes and between different enzymes is far too complex to be unravelled in any detail. Biological systems of extant organisms result from four billion years of trial and error: random mutations and selection constrained by already established structures and functions. If all the mechanisms could be unravelled we would not find a rational design that could be the work of a creator; rather it would give the impression of improvisations and detours reflecting the evolutionary pathways. In a few cases such historical developments can be explained (more or less convincingly) for some details (cf. the evolution of energy metabolism, Chapter 7), but in general there is a very long way to go. A complete sequencing of the genome of an organism does not in itself tell how the organism functions or develops.

The cell membrane

Like DNA, the cell membrane represents continuity from the first cell to all extant cells. DNA cannot produce cell membrane de novo and a pre-existing cell membrane is necessary as a scaffold for producing more membrane.

The cell membrane is built from a double layer of phospholipids (in eubacteria and in eukaryotes) or by (branched) ester bound isoprenoids (in archaebacteria). The fact that lipids spontaneously form such molecular double layers on the water surfaces has not escaped attention and has served as a model for the origin of the cell membrane. The reason for this behaviour is that the linear molecules have a *hydrophobic* and a *hydrophilic* end. 'Hydrophobic' really means that the molecules or parts of a molecule are mutually more attracted than they are attracted to water molecules. The hydrophilic ends of the molecules therefore turn outwards from the membrane (towards the water or air) whereas the hydrophobic ends meet in the middle of the lipid film.

If the lipid film does not cover the entire water surface it is not in its lowest energy state because the hydrophobic ends are exposed to water in the periphery of the film. The lowest energy state would then be obtained if the lipid films form spheres so that all hydrophobic ends are hidden inside the film (Figure 6.1). Such spheres can be made experimentally through appropriate mechanical treatment (shaking, sonication). However, suspensions of spherical lipid films are not always easy to stabilize. In fact, the topic is somewhat complicated and must also take the bending energy of the surface film into consideration. The phenomenon is not entirely unrelated to the coacervates discussed in Chapter 3.

The similarity between these lipid films and cell membranes is so obvious that ideas about the origin of the cell membrane as spontaneously produced lipid spheres would seem to hold some truth. But the model also presents difficulties. Phospholipids are easily formed in the presence of lipids, glycerol, and phosphate. But as discussed

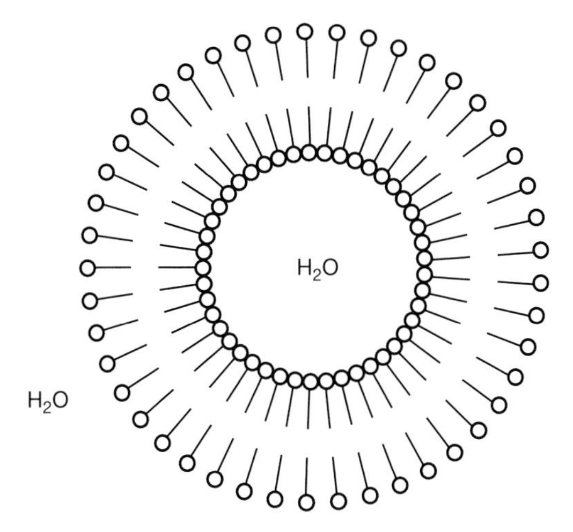

Fig. 6.1 A spherical shell consisting of a double layer of lipid molecules. The hydrophilic ends (with circles) point outwards and towards the inside of the sphere while the hydrophobic ends are hidden inside the membrane.

in Chapter 3 it has proven difficult to synthesize linear fatty acids under assumed prebiotic conditions. Another problem is that the cell membrane of archaebacteria consists of branched isoprenoids. This implies that one of the main groups of organisms at some later point exchanged the bulk material of the cell membranes.

Explaining primitive cell division is also not easy. As we have seen (Chapter 3) coacervates may under some circumstances divide spontaneously. A phospholipid membrane could increase its area by incorporation of phospholipid molecules from solution and divisions could result from mechanical disturbances or from internal growth. The surface of a sphere grows with the two-thirds power of its volume. Assuming a constant addition of mass to the contents of the sphere and to the membrane, then a spherical shape could not be maintained and it would result in budding of the cells.

Neither does the model explain the origin of the cell wall in bacteria. The fact that some bacteria, notably endosymbionts, but also some free-living archaebacteria, do not have a cell wall could be interpreted to mean that the cell wall originated later. In the symbiotic forms, loss of a cell wall is certainly secondary, but it cannot be ruled out that the archaebacteria in question represent an original situation. Eukaryotes did not originally possess a cell wall, but some groups have later developed cell walls (e.g. the cellulose wall of plant cells).

A final difficulty is that a pure lipid membrane only allows small neutral molecules to pass through it. It seems to be difficult to explain simple energy metabolism unless charged molecules could also pass through the membrane.

In all extant organisms the cell membrane is an immensely more complicated structure than just a double lipid membrane and many of the most vital functions of the cell reside in the cell membrane. This is based on a variety of protein molecules

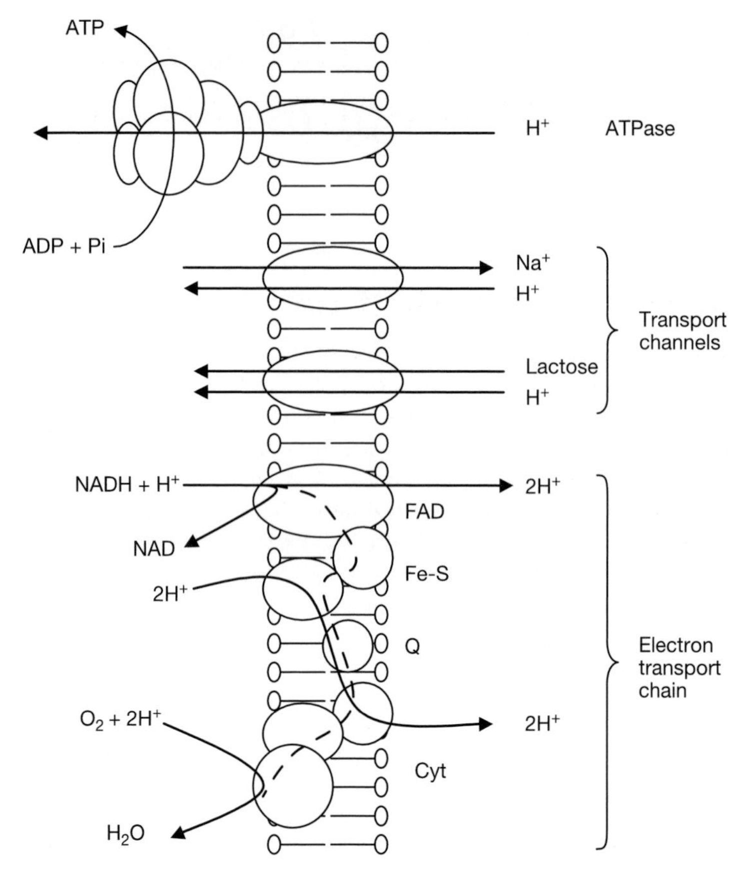

Fig. 6.2 The inner cell membrane of a bacterium, the inside is to the left. Different proteins (including ATPase and transport channels) are imbedded in the membrane. The electron transport chain functions as a proton pump that expels protons from the cell. The path of the electron is indicated by the stippled line. FAD, flavoprotein; Fe-S, iron–sulphur protein; Q, quinone; Cyt, cytochromes among which the last one, cytochrome *c* oxidase catalyses the reduction of O_2 to H_2O. The electron transport chain is explained in more detail in Chapter 7.

that are imbedded in the membrane (Figure 6.2). Most of these penetrate all the way through the membrane and so make contact both to the cytoplasm as well as to the outside world. Energy metabolism (phototransduction, respiration, ATP synthesis) is associated with enzymes embedded in the membrane. Other proteins function as specific transport channels for various solutes (e.g. the sodium pump). Receptor molecules that monitor the concentration of solutes in the environment are also associated with the membrane. Finally, the 'motor' of the bacterial flagellum (not shown in Figure 6.2) is also imbedded in the cell membrane. The membrane proteins are produced in ribosomes that are attached to the inside of the membrane and the complete protein is thereafter inserted in the membrane.

Chapter 7

The evolution of metabolism

Energy metabolism in bacteria

If one could speak of bacteria having a purpose in life it would be to grow and produce two new bacteria. This requires free energy and building materials—the same requirements that must have applied to the first RNA quasi-species as well as to all other kinds of life. The energy always comes from the hydrolysis (splitting with concomitant uptake of H_2O) of energy-rich phosphate bonds. In the case of microorganisms and fast growing other kinds of organisms, by far the greatest energy demand is for the production of polymers (proteins, RNA, DNA) and for active (energy-requiring) uptake of molecules from the surroundings.

The dominating type of high energy phosphate bond is in the form of adenosine triphosphate (ATP) (Figure 7.1) and in special cases other nucleoside triphosphates or inorganic polyphosphates. Polyphosphates result from heating phosphate minerals and may have served as life's first energy source. The process:

$$H_2O + ATP \rightarrow ADP + \text{phosphate}$$

liberates 29.3 kJ per mole (where ADP is adenosine diphosphate). Conversely, when ATP is to be formed from ADP + phosphate this requires more than 29.3 kJ per mole. The mechanisms that are used to generate ATP are referred to as *energy metabolism* (= *catabolic* or *dissimilatory* metabolism). Table 7.1 provides an overview of the basic types of energy metabolism that exist.

Assimilatory metabolism (= *anabolic* metabolism) refers to metabolic processes that serve to build the components of the cell from chemical compounds taken

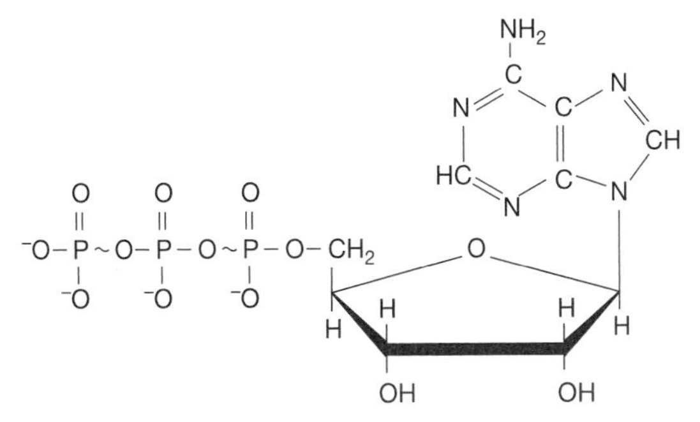

Fig. 7.1 An adenosine triphosphate molecule (ATP). Energy is released through hydrolysis of the high energy phosphate bonds (indicated by ~).

Table 7.1 The fundamental types of energy metabolism

Phototrophy, photosynthesis

Energy source: electromagnetic radiation.

Assimilatory carbon source: CO_2 (in some cases also low molecular weight organic compounds).

Anoxygenic photosynthesis. Electron donors (for the reduction of CO_2) include H_2, H_2S, S^o, and Fe^{2+} (purple and green sulphur bacteria and non-sulphur bacteria; *Heliobacter*).

Oxygenic photosynthesis. Electron donor is H_2O with O_2 as metabolite (cyanobacteria, chloroplasts in algae and plants).

Chemotrophy

Based on reaction between available substrates in the environment. If these are organic compounds it is referred to as *organotrophy*; if substrates are inorganic it is referred to as *chemoautotrophy*. In many organisms the distinction is not quite sharp.

Respiration. The substrate molecules are oxidized with an external electron acceptor.

In *anaerobic respiration* these electron acceptors may be, e.g. NO_3^-, NO_2^-, Mn^{4+}, Fe^{3+}, SO_4^{2-}, and S^o (denitrifiers, iron reducing bacteria, sulphate reducers, etc.).

In *aerobic respiration* the terminal electron acceptor is O_2. Chemoautotrophic bacteria oxidize inorganic compounds such as H_2S, H_2, CH_4, Fe^{2+}, NH_4^+, and Mn^{2+}.

Fermentation. Energy metabolism based solely on the splitting of larger (typically organic) molecules into smaller ones; no external electron acceptor. Many bacteria and some unicellular eukaryotes depend only on fermentation for energy generation. Some respiring organisms (including a few animals) can cope with fermentation alone in the absence of external electron acceptors.

Methanogenesis. Energy metabolism based on the oxidation of H_2 with CO_2 and with CH_4 as metabolite. Some methanogenic bacteria can also ferment acetate or C-1 compounds such as methanol into CH_4.

Methanogenesis is known only in some archaebacteria.

from the environment. The assimilatory metabolism is energy consuming (requires ATP).

Organisms that exclusively utilize inorganic compounds (for example CO_2 as carbon source) are called *autotrophs*. These organisms typically also have an autotrophic energy metabolism based on inorganic compounds or on light energy (*chemoautotrophs* or *phototrophs*). Organisms that depend on organic substrates (typically for energy as well as for assimilatory metabolism) are referred to as *heterotrophs*.

All known types of energy metabolism are represented among the bacteria whereas eukaryotes have a limited repertoire. There are a few assimilatory processes that seem unique to eukaryotes and even to plants and animals, but again the great majority are also known from bacteria. And almost all enzymes known to be involved in energy metabolism are found in one form or another among archaebacteria and eubacteria. They were therefore already present in the universal ancestor.

It has long been known that the processes of energy metabolism are based on coupled redox processes of the type:

$$AH_2 + B \leftrightarrow BH_2 + A$$

where H is hydrogen and where A either reduces or oxidizes B depending on whether the reaction runs from the left to the right, or vice versa. A number of electron or hydrogen carriers are found in cells, and their redox potentials determine which other electron carriers they can oxidize or reduce. One extremely important and universal electron carrier is NAD^+/NADH (oxidized or reduced nicotine adenine dinucleotide); it has a very low redox potential and it is therefore a good electron donor. (In assimilatory metabolism a phosphorylated version NADP/NADPH is mainly used.)

A number of additional electron carriers that are all embedded in the cell membrane, are involved in *respiration* (Figure 6.2). Among these, *flavoprotein* accepts electrons from NADH and delivers them to *Fe-S proteins*. From there the electron passes on to a *quinone* that passes it on to two to three types of *cytochromes*. Eventually the electron is passed on to an external, *terminal electron acceptor*. In aerobic organisms (all animals and plants, but also many kinds of microorganisms) the terminal electron acceptor is O_2 that becomes reduced to H_2O. The last cytochrome (cytochrome *a*) catalyses the reduction of the O_2 molecule and is therefore also referred to as *cytochrome oxidase*. Many kinds of bacteria use other terminal electron acceptors such as sulphate, elementary sulphur, nitrate, ferric iron, or manganic manganese; such cases are referred to as *anaerobic respiration*.

The electron carriers are found in different versions in different organisms, but they are present in all kinds of life. An example is provided by porphyrins (among which the cytochromes belong); its basic structure is shown in Figure 7.2. A metal ion is located in the centre of the molecular structure. In cytochromes this is Cu or Fe and it is the metal ion that can occur in either a reduced or an oxidized form. In all functional porphyrins, the molecule is supplied with side branches and a protein; in addition to the metal it is the side branches and the proteins that determine the particular properties. As shown in Figure 7.2, the molecule has, during evolution, developed different functions that are in some way related to redox processes. The other components of the electron carrier chain involved in respiration (flavoproteins, iron–sulphur proteins, quinones) are also members of molecule families with a common origin for all life forms.

The energy generated by the electron chain derives from the free energy released by the process:

$$\text{NADH} + \text{electron acceptor} \rightarrow \text{NAD} + \text{reduced electron acceptor}$$

(Figure 7.3) and it takes place stepwise. The energy provided is used as a proton pump. Some of the components take up hydrogen, but only deliver an electron to the next step and so the resulting protons (H^+) are pumped out of the cell just as the processes on the inside of the membrane are proton consuming (Figure 6.2). This creates an electrochemical gradient (the exterior of the cell is positively charged and more acid than the inside) somewhat like a rechargeable battery. The electrochemical potential is exploited in different ways by the cell. The protons will, of course, tend to

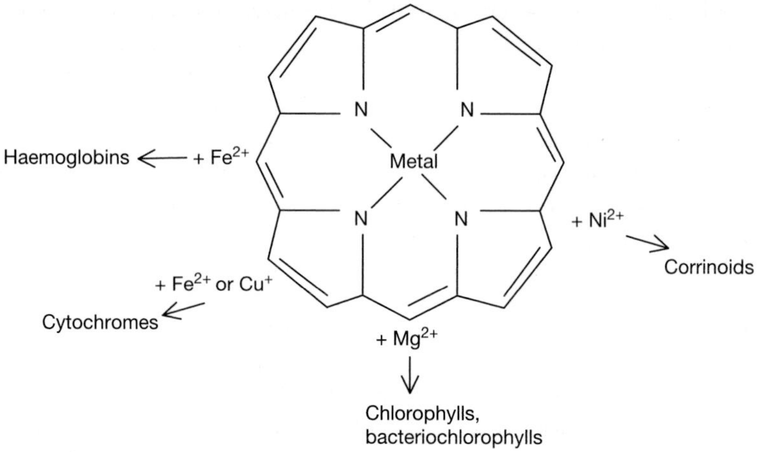

N
N
Metal
N
N
Haemoglobins ← + Fe^{2+}
+ Ni^{2+} → Corrinoids
+ Fe^{2+} or Cu^{+} → Cytochromes
+ Mg^{2+} → Chlorophylls, bacteriochlorophylls

Fig. 7.2 Porphyrin. In modern organisms this molecule structure is associated with a protein and side branches and it contains a metal ion. These determine the precise function of the molecule. During evolution porphyrins have evolved many functions mainly in connection with electron transport. Corrinoids are especially found in methanogenic bacteria. Catalase (that catalyses the dismutation of peroxide into oxygen and water) is also a porphyrin enzyme with Fe.

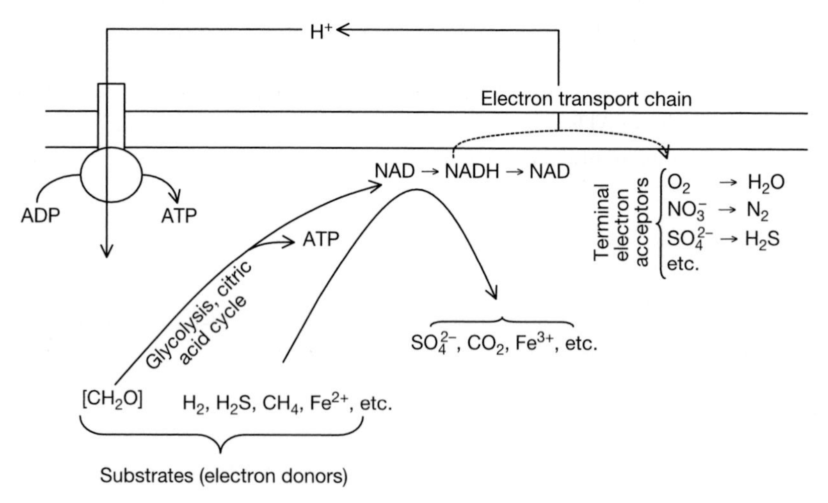

Fig. 7.3 Respiration processes. Substrates reduce NAD to NADH that again is re-oxidized by the electron transport chain (shown in more detail in Figure 6.2). The electron transport chain reduces an external electron acceptor and it functions as a proton pump. The return flux of the protons is coupled to ATP synthesis. When the substrate is carbohydrates these are first degraded by a complex series of processes (the glycolytic pathway and the citric acid cycle) into CO_2 and NADH. This (non-respiratory) fermentative process also generates a small amount of ATP.

migrate back into the cell. This migration is coupled to the production of ATP from ADP + phosphate. ATP generation is catalysed by ATPase: an enzyme complex that is embedded in the cell membrane. The proton flux is also coupled to the energy-requiring transport in or out of the cell (Na^+ and lactose transport are exemplified in Figure 6.2). Also, bacterial flagella that are rigid, rotating helical structures, are driven by the ingoing proton flux. For completeness it should be mentioned that the electron carrier chain of some bacteria pump sodium ions rather than protons out of the cell and ATP production is then coupled to the influx of Na^+.

In addition to an external electron acceptor an external electron donor is also required. This reducing substrate is used for the reduction of NAD to NADH that again feeds hydrogen to the electron carrier chain. Obviously, the substrate must be more reducing than the terminal electron acceptor for the process to run. Possible and realized net respiration processes are shown in Table 7.2. Here the electron acceptors are arranged vertically with a descending tendency to accept electrons and the electron donors (substrates) are arranged horizontally with a descending tendency to donate electrons towards the right. The energetically most efficient processes are therefore placed in the upper left corner of the table. Some of the processes are thermodynamically possible, but are not realized among living organisms. For example no bacteria can oxidize atmospheric nitrogen, because of the high activation energy needed to break the N≡N bond of the N_2 molecule.

When the electron donor is an organic compound (and this is always the case for animals as well as for many kinds of bacteria) it is first necessary to cut the molecules into smaller pieces, and eventually into CO_2 and hydrogen, and the latter then reduces NAD to NADH. This non-respiratory (fermentative) degradation, primarily of carbohydrates, also generates a small amount of ATP (4 moles per mole glucose) whereas the subsequent electron transport chain (with O_2 as terminal electron acceptor) generates 30 moles of ATP per mole of glucose. Some organisms can generate energy only from such fermentative processes and they produce low molecular weight

Table 7.2 Important theoretical and realized respiratory processes in bacteria

	Electron donors							
	$[CH_2O]$	H_2	CH_4	H_2S	Fe^{2+}	NH_4^+	Mn^{2+}	N_2
O_2	+	+	+	+	+	+	+	–
NO_3^-	+	+	–	+	+	+	+	
Mn^{4+}	+	+	–	+	+	–		
Fe^{3+}	+	+	–	+				
SO_4^{2-}	+	+	–					
CO_2	(+)	+						

+ Means that the process is realized in some bacteria, and – means that they are not realized (or undiscovered). Empty spaces indicate thermodynamically impossible (uphill) reactions. The oxidation of organic compounds with CO_2 occurs only in certain fermentative processes, but not in real respiratory processes. Methanogenesis ($CO_2 + 4H_2 \rightarrow CH_4 + 2H_2O$) is formally a respiratory process, but the biochemistry deviates from that of typical respiration.

compounds (lactate, acetate, ethanol, etc.) and H_2 as metabolites. Such fermenting organisms are independent of external electron acceptors, but the energy yield per unit substrate is low. Some animals can survive anaerobic conditions for extended periods of time on the basis of fermentative metabolism. Conversely, there are many bacteria that are partly or totally incapable of degrading organic compounds in conjunction with their energy metabolism.

It is straightforward to explain phototrophy with the background of the above explanation of respiration. Phototrophs have chlorophylls: proteins with a porphyrin centre that includes a magnesium ion (Figure 7.2). Different chlorophylls have different absorption peaks with respect to wavelength: one in the blue part of the spectrum and one in either the red or in the near infrared region. When the chlorophyll molecule is exposed to light it becomes activated and emits an electron. The electrons are taken up by a primary electron acceptor (*phaeophytin*, a chlorophyll without Mg^{2+}). From there the electron is passed on to an electron transport chain, that is basically identical to the one described for respiration, and eventually the electron is returned to the chlorophyll molecule again. The electron transport chain again functions as a proton pump and the return flux of protons is coupled to ATP generation. This process is referred to as *cyclic photophosphorylation.* Phototrophic organisms also need electrons for the reduction of CO_2 to organic matter: that is, for photosynthesis. This requires an external electron donor that is then oxidized in the process. The entire apparatus is called a *photosystem.* Figure 7.4 shows such a system and its function in a green sulphur bacterium. Green sulphur bacteria use only H_2S (or H_2) as electron donor in photosynthesis and produce elementary sulphur (or H_2O) as metabolite. In these bacteria as well as in other photosynthetic organisms the photosystems are associated with membranes. In most forms (but not in green sulphur bacteria) these are invaginations of the cell membrane. Other types of photosynthetic bacteria may use electron donors other than H_2S and H_2, including S^o and Fe^{2+}. These all represent *anoxygenic* photosynthesis because they do not produce O_2 and their chlorophylls are called *bacteriochlorophylls* that absorb near infrared light.

Oxygenic photosynthesis occurs only in *cyanobacteria* (formerly referred to as 'blue-green algae') and in chloroplasts that are descendants of intracellular symbiotic cyanobacteria. It is based on the photosynthetic pigment chlorophyll *a*, uses H_2O as electron donor, and consequently produces O_2 as a metabolite. Oxygenic photosynthesis is somewhat more complicated than anoxygenic photosynthesis in that it involves two photosystems in series. Anoxygenic photosynthesis is surely an evolutionary forerunner for oxygenic photosynthesis. The advent of oxygenic photosynthesis in cyanobacteria, possibly more than 3.5 billion years ago, is one of the most dramatic and significant events in the history of the biosphere, as discussed elsewhere in the book.

It is apparent that respiration and photosynthesis are related processes that to a large extent use the same enzymatic components. The main difference is that in photosynthesis, light activation of a chlorophyll molecule can provide electrons for the electron transport system. In respiration electrons are provided by external electron donors via NADH.

Before we turn to the question of the origin of energy metabolism two additional types should briefly be mentioned. *Methanogenic bacteria* are very important in

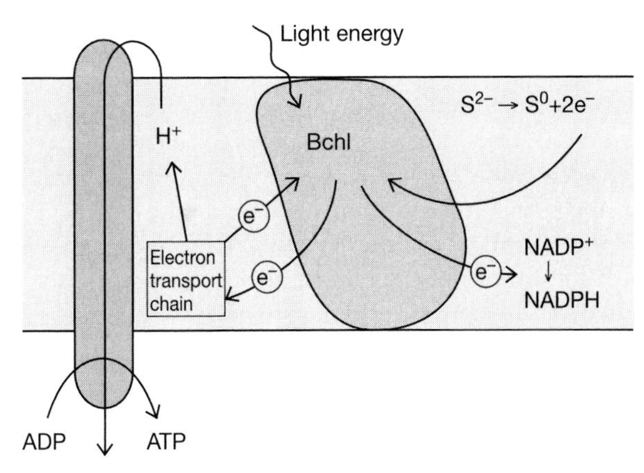

Fig. 7.4 The photosynthetic membrane of a green sulphur bacterium. The light-activated bacteriochlorophyll molecule sends an electron through the electron transport chain (like in respiration) creating a proton gradient and ATP synthesis. The electron eventually returns to the bacteriochlorophyll (cyclic photophosphorylation). If electrons are needed for CO_2 reduction (via reduction of $NADP^+$) an external electron donor is required (sulphide that is oxidized to elemental sulphur).

nature; they belong to the archaebacteria and methanogenesis is only known within this group. Methanogens make a living by oxidizing hydrogen with carbon dioxide and producing methane as metabolite. Formally this is a kind of respiration, but the biochemistry is different (and still not understood in all details). However, the biomolecules involved also show some relationship to those of respiratory organisms. A porphyrin (with nickel) plays a role in the process and, as in respiration, the process also functions as a proton pump with ATP generation coupled to the return flux of H^+.

Finally, a form of energy metabolism has so far been found only in the so-called *extreme halophilic bacteria* that also belong to the archaebacteria. They occur at extremely high salinities. They are the only organisms that live in the south basin of the Dead Sea and they can be grown in saturated NaCl solutions. Normally they live on organic matter that is degraded through aerobic respiration. In light-exposed, oxygen-free habitats these bacteria generate an H^+ gradient (and thus ATP) with the aid of a 'purple membrane'. The cell membrane contains a protein called *bacteriorhodopsin*. In the light these molecules pump protons out of the cells and thus drive ATP synthesis. It appears to be the simplest known type of energy metabolism, but it has not attracted much interest in an evolutionary context. This is probably because it apparently occurs only in this isolated and specialized group of bacteria. However, recent circumstantial evidence suggests that the mechanism also occurs in other archaebacteria that live in oceanic water, and perhaps it will now draw more interest.

In passing: bacteriorhodopsin is chemically closely related to the light-sensitive protein rhodopsin that is found in the retina of the vertebrate eye. Again this illustrates the relatedness of all living things, and the point that all essential biochemistry was developed very early in the history of life.

The earliest evolution of energy metabolism

It was earlier assumed that fermentation of organic substrates represented the most primitive and original form of energy metabolism. Today most people concerned with this sort of question would tend to assume that some simple form of membrane-associated chemo- or photoautotrophy represents the original form of energy metabolism. There are several reasons for this. First, a large number of fermentative bacteria have been shown to be descended from photosynthetic or respiratory bacteria through loss of enzymatic machinery. Also, fermentation is not a simple process (as discussed in the following section) since it requires a long list of enzymes.

It can be assumed that the first membrane-bound electron transport mechanisms were based on the simple functional molecules that also represent the functional units of modern enzymes, but without the protein component. The protein components would be a later development that improved efficiency and specificity, but they are not essential for the basic functional role. One might imagine that such 'naked' molecules such as quinone, metal-containing porphyrins, and inorganic Fe-S molecules would be common in an anoxic, prebiotic Earth. Such molecules could have become incorporated into the primitive cell membrane and so be responsible for a primitive electron transport system (cf. discussion on coacervates in Chapter 3) or a kind of photochemical energy transduction.

Figure 7.5 shows some hypothetical examples of a primitive form of energy metabolism. The simplest, perhaps functional, model (Figure 7.5, left) depends only on a Fe-S complex, a quinone (Q), and a porphyrin molecule (P) that can be photoactivated. An example of such a simple system could function as a light-driven proton pump. Through the catalytic oxidation of an external electron donor this would also provide electrons for CO_2 reduction. Such a primitive reaction could be the production of pyrite + hydrogen from the reaction of ferrous iron and sulphide as suggested by Wächterhäuser (Chapter 3) in connection with his 'pyrite-world'. A slightly more advanced stage (Figure 7.5, right) is quite close to modern photosynthesis: the photosensitive porphyrin has become a protochlorophyll (Chl) and another porphyrin, a precursor for a cytochrome (Cyt) provides a mechanism by which electrons can be channelled back to chlorophyll allowing for cyclic photophosphorylation.

It is a convincing argument that the electron transport system with its essential components (Fe-S, quinones, cytochromes) is very ancient because the components occur almost universally in all of the main groups of eubacteria and archaebacteria, and when absent there is evidence to show that this is a secondary loss. There is also evidence (but here there is no unanimous agreement) that respiring organisms are derived from phototrophic ones through secondary loss of chlorophylls. Presumably electromagnetic radiation was the primary energy source for the generation of organic molecules and for the first approaches to life—just as light is the principle energy source that drives the extant biosphere. The universal phylogenetic tree (Chapter 12) shows that unrelated groups of photosynthetic eubacteria have 'deep roots' and that representatives of many groups that include photosynthetic forms have secondarily

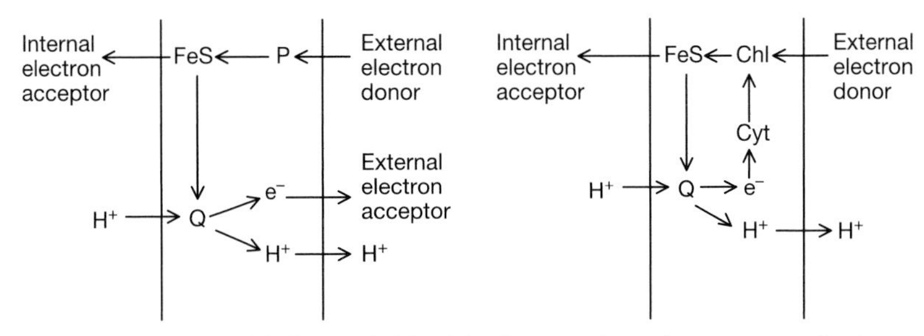

Fig. 7.5 Theoretical models for a primitive (photic or respiratory) energy transduction apparatus built into the cell membrane. FeS, iron sulphide; P, a porphyrin molecule; Q, quinone; Chl, a protochlorophyll (a photo-excitable porphyrin); Cyt, cytochrome. For further explanation, see text. Idea after B. K. Pierson and J. M. Olson. (1989). Evolution of photosynthesis in anoxygenic photosynthetic prokaryotes. In *Microbial mats* (ed. Y. Cohen and E. Rosenberg), pp. 402–27. American Society of Microbiology, Washington DC.

lost chlorophylls and have become dependent on external chemical reductants for respiration processes.

A group within the eubacteria, the *proteobacteria*, provides a good example of the evolution from photosynthetic forms to species that depend on respiratory metabolism. Most subgroups include photosynthetic forms (the so-called *purple sulphur bacteria* and the *purple non-sulphur bacteria*) and it would be difficult to imagine that very similar types of photosynthesis had evolved independently on several occasions. But all the subgroups also include forms with anaerobic or anaerobic respiration.

Figure 7.6 provides a schematic presentation of photosynthesis and respiration in purple non-sulphur bacteria. The name for these bacteria is clumsy and also somewhat misleading in that they are capable of using H_2S in addition to H_2 in photosynthesis. These purple or brownish bacteria live in anaerobic or oxygen-poor, but illuminated habitats such as the surface layers of shallow marine sediments. They have a broad metabolic repertoire. In the light they prefer complete anoxia and carry out anoxygenic photosynthesis preferably with hydrogen as electron donor. In the dark they prefer habitats with a low O_2 content and their energy requirements are covered by aerobic respiration. They use the same electron transport system in both cases. It is evident that if they were to lose their bacteriochlorophyll they would be able to carry on as respiratory bacteria as long as they have access to the necessary substrates (low molecular weight compounds such as lactate and acetate, H_2) to generate reduced NADH. This form of evolution has apparently taken place many times. Indeed, intermediate forms are found today. The recently discovered *aerobic anoxygenic phototrophs* are basically oxygen respirers that use organic substrates. However, they contain some bacteriochlorophyll and, when starved, they can still generate some ATP through phototrophy, but they cannot sustain growth on the basis of light.

The purple non-sulphur bacteria are particularly interesting because the electron transport system in these bacteria is almost identical to that of mitochondria.

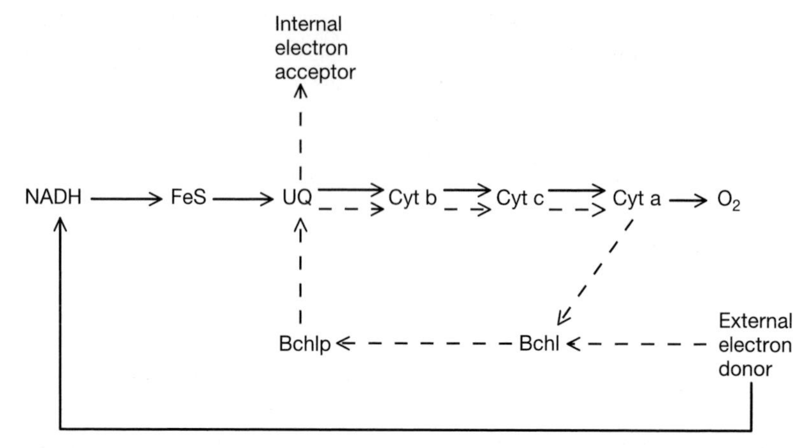

Fig. 7.6 The photosynthetic and respiratory system in a purple non-sulphur bacterium. Stippled line, path of electrons in light (cyclic photophosphorylation, photosynthesis); uninterrupted line, in the dark (respiration). FeS, iron–sulphur protein; UQ, ubiquinone; Cyt, cytochromes; Bchl, bacteriochlorophyll *a*; Bchlp, bacteriophaeophytin. Loss of bacteriochlorophyll results in a non-photosynthetic, respiring organism.

Many other groups of respiring and phototrophic bacteria show some deviations from this. This accords with the well-established theory that mitochondria are descendants of endosymbiotic purple non-sulphur bacteria (Chapter 8) or at least from species belonging to the group (the α-proteobacteria) that also includes the purple non-sulphur bacteria.

Our present knowledge suggests that a light-driven energy metabolism, somewhat resembling that depicted in Figure 7.5, represents a very early form of energy metabolism. As in so many other cases there are difficulties: no one has so far suggested on how proton flux could originally be coupled to phosphorylation as an energy conservation without the complicated ATPase enzyme.

Methanogenesis (from $CO_2 + H_2$) is generally (and probably correctly) believed to be a very ancient type of energy metabolism, but there are no simple suggestions how methanogenesis could have developed from scratch since it deviates so much from other types of energy metabolism and in part involves special enzymes.

Some general considerations on assimilatory metabolism—the origin of carbohydrate catalysis and fermentation

The number and complexity of assimilatory and synthetic pathways are considerable and cannot be discussed in any detail here. However, in spite of this apparent diversity it is also based on the differential use of a much more limited number of components.

It is difficult to imagine how complicated biochemical processes involving many steps and cycles could possibly have originated (cf. Figure 7.8). But it is somewhat easier in the case of assimilatory than of dissimilatory processes. It can be imagined that a proto-organism used some sort of essential component *A* that originally occurred in the primordial soup. Eventually, however, *A* became depleted because it was no longer synthesized abiotically or because demand was higher than supply. It is then possible that some mutation allowed the organism to convert a related compound *B*, that was still available from the environment, into *A*. As *B* became unavailable the organism had to 'learn' to produce *B* from *C*. In this way the organism had developed a biochemical pathway:

$$C \rightarrow B \rightarrow A$$

that could be extended to *D*, etc.

Another mechanism that would contribute to explain the biochemical complexity is the fact that such pathways can also work in reverse. All metabolic processes are, in fact, reversible and a given enzyme may catalyse the process in either direction. The direction of a process, or of several coupled processes, is determined by the addition of a substrate at one end and the removal of the product at the other end. The process $A \leftrightarrow B$ will proceed to the right if *A* is added and *B* removed and vice versa. As an example we may think of the assimilatory reduction of CO_2 (Figure 7.7, top). Such a process will require energy (use ATP) and it will be reductive (require H_2 in the form of reduced NADH or some other cellular electron donor). The process might, however, also run the opposite way and become a catabolic (dissimilatory) and oxidative pathway, degrading and oxidizing organic matter into CO_2 and preserving the free energy gained in the form of ATP.

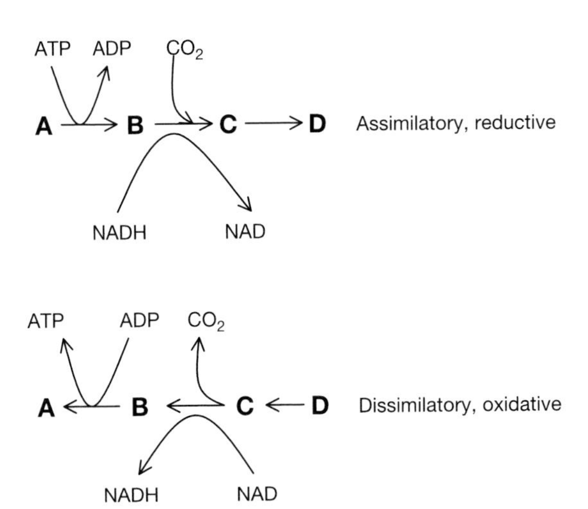

Fig. 7.7 A series of coupled biochemical reactions may be assimilatory and reductive, that is reducing CO_2 to organic matter at the expense of ATP (top) or it may be reversed, becoming dissimilatory and oxidative and thus generate ATP (bottom). The origin of fermentative pathways (bottom) is easiest to understand as an assimilatory pathway running in reverse.

The origin of the *glycolytic pathway*, *fermentation*, and the *citric acid cycle* may be explained in this way. The complete system in an aerobic bacterium is shown in Figure 7.8. It consists of two main components. The glycolytic pathway uses carbohydrates as substrate; it is an anaerobic degradation of a glucose molecule (six C atoms) into two pyruvate molecules (three C each). This process provides a net gain of two moles of ATP and two moles of reduced NADH. The glycolytic pathway is simplified in Figure 7.8; in reality it includes seven intermediate products and a similar number of enzymes. Fermenting organisms may continue to ferment pyruvate into a variety of metabolites (acetate, butyrate, lactate, ethanol, hydrogen) with a further gain of one to two ATP.

In aerobic organisms pyruvate is decarboxylated (removal of CO_2) to acetyl coenzyme *A* (acetate coupled to a coenzyme) and incorporated into the citric acid cycle. In principle the citric acid cycle is a cyclic process in which two C atoms are oxidized to CO_2 for each turn of the cycle. This oxidation results in the production of

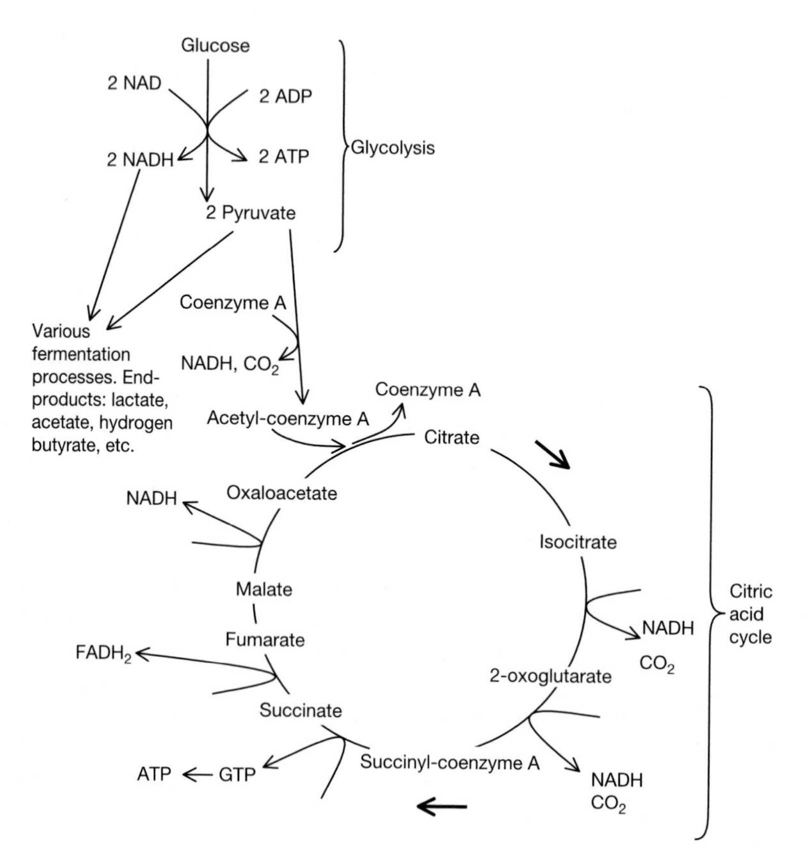

Fig. 7.8 Glycolysis and the citric acid cycle serve to degrade glucose to CO_2 and H (in the form of NADH and $FADH_2$) that is then oxidized through respiration. Fermenting organisms use only the glycolytic pathway plus a few supplementary processes in order to generate a modest amount of ATP.

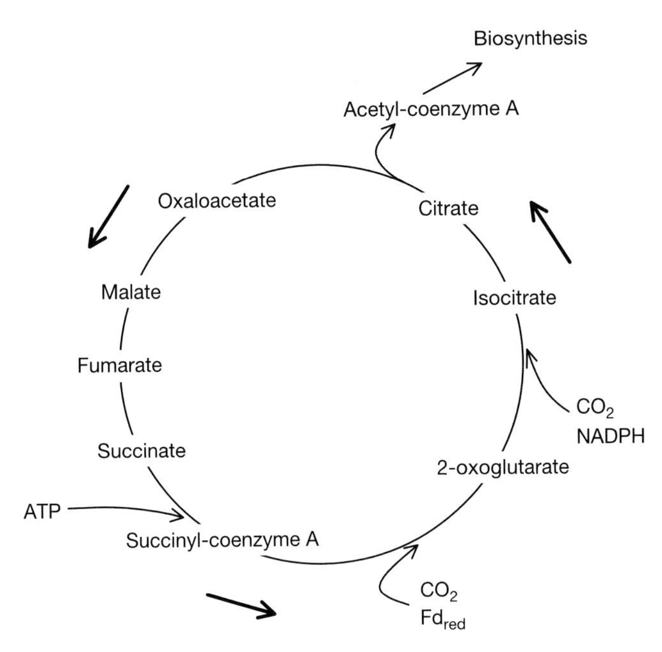

Fig. 7.9 In green sulphur bacteria and in some archaebacteria a reverse citric acid cycle is used for the assimilation of CO_2. It must be assumed that this was the original function of the citric acid cycle, that only secondarily took over the role as a dissimilatory and oxidative process for the degradation of organic matter.

four molecules of H_2 bound as NADH and $FADH_2$ (flavin adenine dinucleotide, another cellular electron carrier). Together with the NADH produced through glycolysis, these are then oxidized via the electron transport system in the cell membrane (Figures 6.2 and 7.3). The entire system is also found in most eukaryotes; here glycolysis takes place in the cytoplasm, the citric acid cycle in the mitochondrial matrix, and the electron transport in the mitochondrial inner membrane (Chapter 8).

The origin of the glycolytic pathway and the citric acid cycle is difficult to understand, but easier if we consider it as an assimilatory process running in reverse. There are several different ways in which CO_2 can be assimilated into organic matter. Cyanobacteria (and thus plants and algae) and photosynthetic purple bacteria use the so-called Calvin cycle. But green sulphur bacteria and some archaebacteria use a reversed citric acid cycle (Figure 7.9). Here the cycle is reductive, it assimilates CO_2, it is ATP requiring, and its function is to produce acetate from CO_2 for further biosynthesis. There is little doubt that this was the original function of the citric acid cycle and as such it is easier to understand its origin.

Syntrophy

In the next chapter we will leave the bacteria for a while. But before this we will shortly discuss interactions between different kinds of bacteria. Bacteria cannot take up particulate matter or macromolecules directly, only low molecular weight compounds can pass through their cell membrane (single-stranded DNA being one exception). In the first half of the history of the biosphere, when only bacteria were around, no one ate each other and this innocent situation lasted until the origin of the eukaryotes. Endosymbiosis, which played such an important role in the origin and evolution of eukaryotic cell, was also not possible in a purely bacterial world.

This is not to say that different physiological types (species) of bacteria do not interact in nature or form biotic communities. Bacteria interact in a multitude of ways and bacterial community ecology may be complex and interesting. For example, bacteria compete for common resources and some carry out chemical warfare (antibiotics) to kill or inhibit competitors.

An interesting type of interaction is based on the fact that some bacteria depend on the metabolites of other bacteria for a living. Anaerobic bacteria in particular are very specialized in this respect. Thus no single physiological type of bacteria can completely degrade organic compounds such as carbohydrates into CO_2 and CH_4 or into CO_2 and H_2S; this requires a consortium of two or more different species. Bacteria with anaerobic respiration lack the enzymes necessary for glycolysis and are often also not in the possession of a citric acid cycle. Methanogenic bacteria can utilize only very few substrates, primarily H_2. They therefore depend on fermenting bacteria to degrade carbohydrates into low molecular weight compounds and H_2.

The relation between hydrogen-producing (fermenting) bacteria and hydrogen-consuming bacteria provides an example of *syntrophy*. The energetically most efficient type of fermentation of carbohydrates is into acetate + CO_2 + H_2. But this process can proceed only if the ambient H_2 pressure is extremely low, otherwise it becomes a thermodynamic uphill process. There are, for example, bacteria that can ferment ethanol into acetate + H_2, but it comes to a halt if the H_2 tension increases within the cell. Thus it requires the presence of another, H_2-consuming bacterium in the immediate vicinity—in Figure 7.10 exemplified by a methanogen. The methanogen thus receives its substrate from the fermenting bacteria that in turn is kept going by the immediate removal of H_2. There are many examples of such

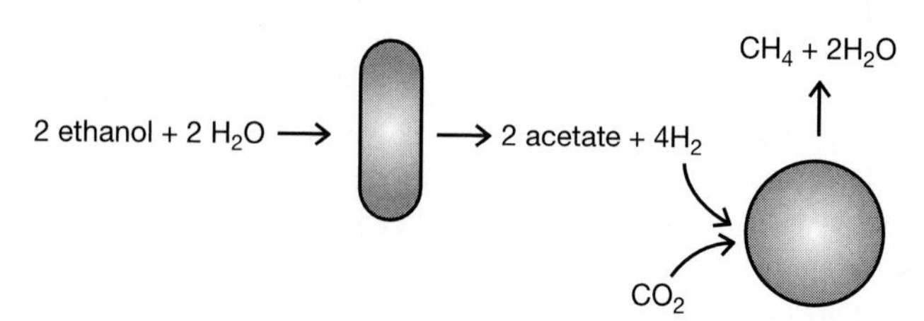

Fig. 7.10 Syntrophy: two bacteria with complementary types of metabolism are coupled. One of the bacteria is fermenting ethanol to acetate and H_2 and the other is a methanogen that makes CH_4 from H_2 and CO_2. The metabolism of the fermenting species requires a low H_2 tension and the other species depend on access to H_2. Such syntrophic pairs must live in close physical contact and the dynamics of the system implies that their growth rates are coupled.

syntrophic pairs of physiologically complementary bacteria. It can be shown that the growth and reproduction of the partners become coupled and that the common growth rate corresponds to the net energy yield of the process:

$$\text{ethanol} \rightarrow CH_4 + \text{acetate}.$$

The example shown in Figure 7.10 was originally described as a pure culture of one kind of bacterium with the ability to ferment ethanol into methane and acetate. Only later was it found that the culture consisted of two different, but quite interdependent species.

Chapter 8

The eukaryotic cell

Properties of the eukaryotic cell

The distinction between prokaryotes ('before nuclei') and eukaryotes ('with nuclei') was made by the French zoologist É. Chatton around 1930. A more systematic analysis of the fundamental differences between pro- and eukaryotes was provided by R. Stanier about 30 years later.

Eukaryotes include all organisms that are not bacteria (prokaryotes = eubacteria + archaebacteria). The eukaryotes include the three large groups of multicellular organisms. That is, animals, (vascular) plants, and the higher fungi plus a large number of rather unrelated groups of unicellular organisms often referred to as protozoa, protophytes (unicellular algae), multicellular algae, and some fungi-like organisms. Eukaryotes that belong to neither the animals, plants, or fungi are often collectively referred to as *protists.*

The eukaryotes represent an apparent discontinuous jump in complexity relative to the prokaryotes and no convincing missing link has been found. The so-called endosymbiosis theory for the origin of the eukaryotes can, in a satisfactory manner explain the origin of some characteristic eukaryote organelles: mitochondria, chloroplasts, and perhaps peroxisomes. These represent important characteristics of eukaryotes, but there are many additional problems when trying to derive eukaryotes from prokaryotes.

It is commonly stated that the eukaryotes arose 1.5–2 billion years ago. This accords with the age of the earliest fossils that have been interpreted as remains of unicellular eukaryotes (Chapter 13) and with estimates for the time of the origin of the mitochondrion as based on gene sequences (Chapter 12). The origin of the mitochondrion, and thus of aerobic energy metabolism in eukaryotes, is a significant event and was a pre-condition for their further evolution and diversification (Chapter 14). Nevertheless, molecular trees (Chapter 12) suggest that the eukaryotes have 'deeper roots'. Examples of unicellular eukaryotes are shown in Figures 8.3 and 8.4 (see also Plates 2 and 3).

Eukaryotes possess a number of special characteristics (Table 8.1 and Figures 8.1 and 8.2). The term eukaryotes refers to the presence of a membrane-enclosed nucleus. It contains two or more linear chromosomes and their ends are capped by *telomeres.* A special type of protein, *histone*, is associated with the DNA of the chromosomes. In contrast to the prokaryotes, replication can take place simultaneously at several sites on each chromosome, and it has been proposed that the reason why eukaryotes have more DNA than prokaryotes is that replication is faster. Another peculiarity of the genome in eukaryotes is the presence of *introns*: pieces

Table 8.1 Important differences between eukaryotes and prokaryotes

	Eukaryotes	**Prokaryotes (eubacteria + archaebacteria)**
Nucleus + nuclear membrane	+	–
Chromosomes	>1, linear, DNA bound to histones	One circular chromosome
Sex, meiosis	+ (–)	–
Cell division	Mitosis	No mitosis
Introns, repetitive DNA	+	– (Introns occur in archaebacteria)
Ribosomes	80S	70S
Cytoskeleton (actin and tubulin filaments)	+	–
Phagocytosis, pinocytosis	+	–
Motility	Eukaryotic flagella, amoeboid motion	Prokaryotic flagella, gliding motility
Internal membranes	+	In some groups
Membrane-covered organelles (mitochondria, chloroplasts, peroxisomes)	+ (Chloroplasts only in photosynthetic forms, a few primitive groups also lack mitochondria and peroxisomes)	–
Steroles in cell membranes	+	–
Sizes	2–1000 μm (typically around 10–100 μm); some are larger	Typically about 1 μm (0.5–50 μm)

of DNA that intersect genes. Introns are not translated because the mRNA undergoes a maturation process in which the introns are cut out and the gene is spliced together again into an informative sequence prior to translation. Introns are, however, also found in archaebacteria. Eukaryotes also carry a large, but varying amount of *repetitive DNA* that is not transcribed. Eukaryotes have substantially more DNA than prokaryotes, but a large part of it is not translated and does not seem to have any other function. In many eukaryotes such *junk DNA* constitutes the largest part of the genome. It is now generally believed that it represents parasitic DNA or 'selfish genes' that can hang on as long as the fitness of the carrier is not severely affected.

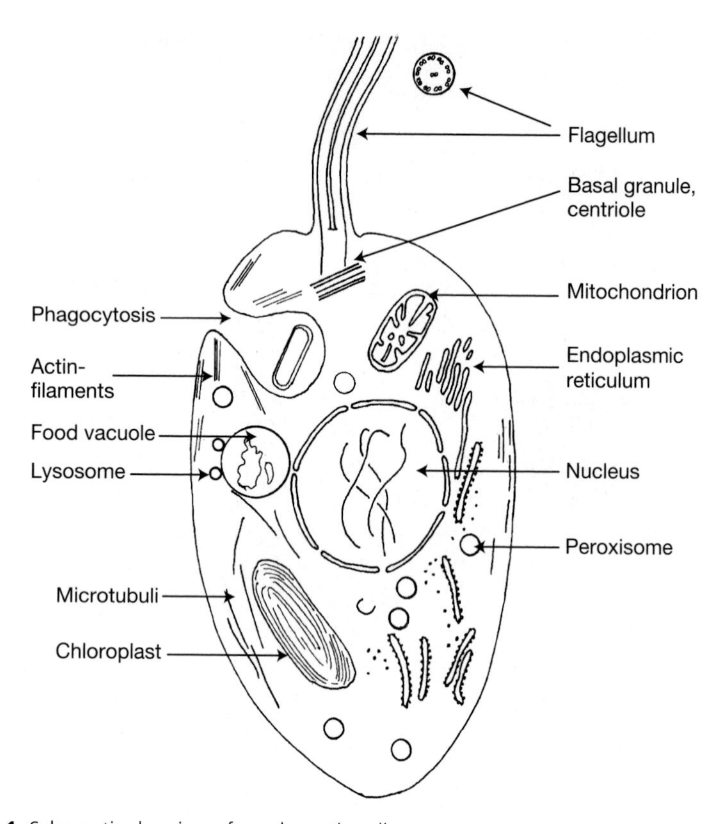

Fig. 8.1 Schematic drawing of a eukaryotic cell.

The ribosomes of eukaryotes are larger than those of prokaryotes. The large and small components are referred to as 80S and 18S in eukaryotes; for prokaryotes the two components are called 70S and 16S, respectively (S is a unit for molecular size as measured by the sedimentation speed in a centrifuge).

Reproduction of eukaryotic cells takes place through mitosis, a process ensuring that each daughter cell receives a complete copy of the genetic information. Haplo-diplont cycles and sexual processes are other characteristic features of eukaryotes that will be discussed in Chapter 10.

The replication of the chromosomes takes place prior to mitosis, they then appear as doubled-stranded, and the two *chromatids* are attached only at one site, the *centromere*. During the cell division the centromeres divide and the resulting two chromosomes are pulled apart to end up in their respective daughter cells.

Ribosomes are found in the cytoplasm outside of the nucleus and translation is therefore spatially isolated from transcription. The mRNA strands pass out to the ribosomes via pores in the nuclear membrane.

Perhaps the single most important characteristic of eukaryotes is the presence of a cytoskeleton. It consists of two types of protein filaments. *Actin filaments* consist of

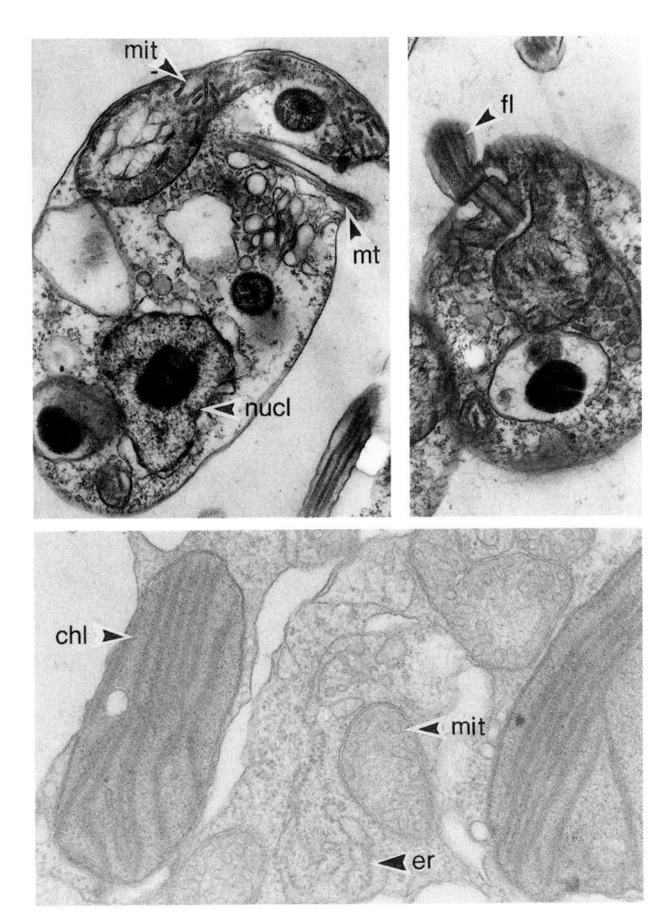

Fig. 8.2 Structures in eukaryotic cells. (Top) Two sections of the approximately 5 μm long flagellate *Bodo jaculans* as seen in the electron microscope. nucl, cell nucleus; mit, mitochondrion; mt, microtubuli associated with the 'mouth' of the cell through which food particles (bacteria) are ingested; fl, flagellum. The centriole/basal body is seen beneath the flagellum. This flagellate belongs to a group (Kinetoplastida) in which the mitochondria contain an unusually large amount of DNA (seen as filaments in the swollen part of the mitochondrion). The function of this is unknown. (Original) (Bottom) Part of the dinoflagellate *Gymnodinium fusus* at high magnification. chl, chloroplast with thylakoids; mit, mitochondria with double membrane and cristae; er, endoplasmic reticulum with associated ribosomes. Unattached ribosomes are also seen. (J. Larsen)

the protein actin. *Tubulin filaments* (*microtubules*) have a larger diameter; they are hollow cylinders composed of two versions of the protein tubulin. Both types of filaments serve as a sort of skeleton that supports the cell mechanically. They also serve for cellular motility whether intracellular (such as moving organelles around in the cell and during cell division) or for swimming or creeping around using flagella (cilia) or amoeboid motion. The motility is based on sliding of parallel filaments relative to one another. This is accomplished by special motor proteins (myosin in the case of actin filaments and dynein in the case of microtubules) and with ATP as energy source. Animal muscle cells are based on actin–myosin and they are filled with bundles of parallel actin filaments.

The cytoskeleton is important because it allows for advective transport inside the cell and this allowed for the evolution of larger cells; the maximum size attained by eukaryotic cells is about 1 mm. The cytoskeleton also allows for a much more varied cell shape, as a rigid cell wall is not necessary to hold the cell together. Finally the cytoskeleton allows for the uptake of particulate matter and dissolved macromolecules from the surroundings. This takes place through an invagination of the cell membrane that forms a food vacuole in which the food is digested. The process is called *phagocytosis* in the case of particles and *pinocytosis* in the case of large molecules; collectively the processes are referred to as *endocytosis.*

The *centriole* is a peculiar organelle that is built from microtubules. It consists of two perpendicularly oriented cylinders each built from nine short pairs of microtubules arranged in parallel. During cell division (mitosis) one-half centriole moves to the opposite pole of the cell. After dissolution of the nuclear membrane and division of the chromosomes microtubules grow out from the centrioles and attach to a centromere on each chromosome. Through these microtubular bundles the chromosomes are arranged in the equatorial plane of the cell and then pulled apart. This secures the correct distribution of chromosomes between daughter cells. It is still unclear whether the centrioles are 'replicators' in the sense that they eventually divide to produce a double set of microtubular cylinders or whether the new cylinder arises only in the vicinity of the old one.

The eukaryote flagellum is an organelle that is related to the centriole. Actually, the basis of the flagellum (the *basal body* or *kinetosome*) is a sort of centriole and in some protists the basal body also functions as a centriole (Figures 8.1 and 8.2). The flagellum is surrounded by an extension of the cell membrane. It is distinguished from the centriole in that it, in addition to the nine double microtubules, also contains two central microtubules. The sliding relative to one another of the peripheral microtubule pairs results in motility. Some forms (e.g. ciliates) carry numerous flagella; these are then usually called cilia.

Almost all eukaryotic cells are characterized by internal membrane systems that form a complex of cisterns and channels (the endoplasmic reticulum). These membranes may be smooth or covered by ribosomes. *Lysosomes* are membrane-covered vesicles that derive from the endoplasmic reticulum; they contain hydrolytic enzymes that are emptied into food vacuoles in order to digest the food. There are a few, presumably primitive types of eukaryotes that are devoid of such internal membrane systems and these organisms also lack mitochondria and peroxisomes that are otherwise universally found in eukaryotes.

Mitochondria are surrounded by a double membrane, and the inner membrane carries invaginations (*cristae*) that may fill a large part of the volume (Figure 8.2). The main function of the mitochondrion is energy metabolism. Mitochondria have their own chromosome of bacterial-type although it codes only for their structural RNA and for a minor part of the necessary proteins. Mitochondria reproduce themselves by division. Chloroplasts share many of these properties. They too have their own genome and reproduce by division. Chloroplasts are surrounded by two or more membranes and they contain stacked photosynthetic membranes, *thylakoids* (see Figure 8.2).

Peroxisomes are spherical organelles surrounded by a single membrane. They are filled with enzymes and especially peroxidases that catalyse the dismutation of hydrogen peroxide into oxygen and water. They are part of the protection against oxygen toxicity (Chapter 11). Peroxisomes do not have a genome, but appear to multiply by divisions.

This sketchy description of eukaryotic cells suffices to see that they represent a leap in complexity when compared with prokaryotes. Apparently, eukaryotes possess a number of traits that have no counterpart among the bacteria.

The origin of the eukaryotic cell

There are no organisms that can be considered as intermediates between prokaryotes and eukaryotes. A small group of flagellated or amoeboid organisms are characterized by the absence of membrane-covered organelles such as mitochondria and of an endoplasmic reticulum. These organisms also branch off early from the remaining eukaryotes in phylogenetic trees (Chapter 12). But in spite of these, perhaps primitive, traits, they do after all possess most characteristics of eukaryotes: a cytoskeleton, eukaryotic flagella, and a nucleus with chromosomes.

In certain respects the eukaryotes seem more related to the archaebacteria than to the eubacteria. This applies to the nucleotide sequences for some genes (mainly those that are involved in the genetic apparatus and in protein synthesis), the presence of introns in the genome, and in certain details of the protein synthesis (explaining why some antibiotics affect eubacteria, but not archaebacteria and eukaryotes). On the other hand, there are genes (especially those involved in energy metabolism) that are more closely related to those of eubacteria, and the cell membrane of eukaryotes is similar to that of eubacteria.

Among the archaebacteria it is especially the *thermophilic archaebacteria* that have drawn interest in this context. These organisms are known from hot springs and other geothermal phenomena. They are characteristic in not possessing a cell wall and some have structures that resemble a cytoskeleton. Some look a bit amoeba-like such as, for example, *Thermoplasma*. The thermophilic archaebacteria have different kinds of energy metabolism, mostly based on sulphur compounds using sulphate or sulphur to oxidize organic matter, or they make a living by oxidizing sulphide with O_2. Forms that can use Fe^{3+} or Mn^{4+} as oxidants in respiration also occur. The mentioned *Thermoplasma* can also thrive on organic matter with aerobic respiration. It is often assumed that the original eukaryotes, before they acquired mitochondria, had a purely fermentative metabolism. In the universal phylogenetic tree (Chapter 12) the eukaryotes seem to be somewhat closer to the thermophilic archaebacteria than to other types of prokaryotes. But after all, these archaebacteria do not possess many traits that can explain the origin of eukaryotes.

One suggestion that could explain why some eukaryotic genes resemble those of archaebacteria, and others resemble those of eubacteria, as well as the apparent leap in structural complexity, would be that the eukaryotes originated as a kind of chimera between an archaebacterium and a eubacterium (this idea should not be mixed up with the more well-documented theory for the origin of mitochondria and chloroplasts). Thus the complex could originally have been an archaebacterium that somehow managed to reside inside a eubacterial cell. Eventually, the genes of the eubacterial host moved to the chromosome(s) of the archaebacterium by horizontal gene transfer and the archaebacterial symbiont ended up as the nucleus. Other models for the origin of eukaryotes have implied various topological twisting of bacterial cells to explain how the nuclear membrane and mitosis arose. All these proposals may hold some truth, but they do not, for the time being, constitute testable scientific hypotheses.

Altogether there are several important eukaryotic traits that cannot be satisfactorily explained. Recently, tubulin- and actin-related fibrils have been found in some bacteria, but not motor proteins. But satisfactory explanations for the structure of eukaryotic chromosomes or for the origin of mitosis are still lacking. It is, of course, still possible that some sort of missing link will be found; after all, many new types of microbes have been found during the last few decades.

The origin of mitochondria and chloroplasts

It is not a new idea that chloroplasts of plants and algae represent endosymbiotic cyanobacteria: originally it was suggested in 1910 by the Russian biologist Mereschowsky. A couple of decades later the American biologist Wallin suggested that mitochondria (the function of which was then unknown) were endosymbiotic bacteria. This idea fell (reasonably enough) into discredit after Wallin claimed that he could grow mitochondria as bacteria, and the idea was largely forgotten. In the 1960s Lynn Margulis revived the idea of organelle origin as endosymbiosis. Originally met with some scepticism, this model for the origin of mitochondria and chloroplasts is now well documented and generally accepted.

Both types of organelles reproduce by division (are replicators), have their own bacteria-type chromosome, an apparatus for transcription and translation, and bacteria-type ribosomes. Evolutionary trees based on the sequence of rRNA genes (Chapter 12) unambiguously place mitochondria and chloroplasts among the α-group of the proteobacteria, and the cyanobacteria, respectively. Finally, there are many recent examples of endosymbiotic prokaryotes inside eukaryotic cells with a functional significance that is more or less similar to that of mitochondria and chloroplasts.

Mitochondria are surrounded by a double membrane. There are different interpretations of this feature; the most likely one is that the two membranes represent the inner and outer membrane of bacteria. The inner membrane is invaginated forming tubular or discoid cristae. The mitochondria import pyruvate and (indirectly) NADH deriving from glycolysis. In the mitochondrial matrix, pyruvate is oxidized to CO_2 through the citric acid cycle (Chapter 7, Figure 7.8). The resulting NADH is then oxidized by the electron transport chain with O_2. The electron transport chain resides in the inner membrane of the mitochondrion; its biochemistry is very similar to that of purple non-sulphur bacteria (Figure 7.6).

The theory implies that the original eukaryote was an anaerobic fermenting organism and that an aerobic bacterium established itself as a symbiont within this protoeukaryote. The symbiont could then utilize metabolites of the host for its own aerobic metabolism. It is possible that the original symbiont was photosynthetic, like modern purple non-sulphur bacteria and that it later lost its capability of photosynthesis.

The chromosome of modern mitochondria codes for only about 1% of the proteins they need; the remaining genes have moved to the chromosomes of the host that consequently carry a number of typical eubacterial genes. This has necessitated that the mitochondrion imports several proteins for which special mechanisms have evolved. Structurally there appears to be two kinds of mitochondria: those with tubular cristae (occurring in most protist groups) and those with discoid cristae (a few groups of unicellular protists and in animals, plants, and fungi). It is generally believed, however, that mitochondria arose only once.

Mitochondria are absent in only a few types of anaerobic protists. Until recently this was considered as a primitive trait, but a mitochondrial gene has now been found in the chromosome of one of these groups, and so it is possible that the absence of mitochondria is a secondary development. Other protists groups have also become secondary anaerobes. These include some ciliate and flagellate groups and the fungi-related chytrids. They have retained organelles with a mitochondrial structure (hydrogenosomes) that serve for further fermentation of pyruvate into acetate and H_2 with an additional yield of ATP.

The function of chloroplasts is oxygenic photosynthesis. Chloroplasts do not occur in all eukaryotic groups and some have secondarily lost them. Chloroplasts have presumably evolved independently at several occasions and altogether the history of chloroplasts is complex. This is supported by the frequency by which photosynthetic endosymbionts occur among various extant eukaryotic groups. In some cases (e.g. in dinoflagellates and cryptomonad flagellates) the chloroplast represents a kind of second-generation endosymbiont. Here the chloroplast represents the remains of an endosymbiotic eukaryotic phototroph. Also in these cases a horizontal gene transfer from endosymbiont to the host nucleus has taken place. But in cryptomonad chloroplasts there are still remains of a eukaryotic nucleus. This type of double-symbiosis results in several sets of membranes surrounding the chloroplast in which the outermost membrane represents the cell membrane of the latest endosymbiont.

The principle photosynthetic pigment of chloroplasts (and cyanobacteria) is chlorophyll *a*. In addition, some other chlorophylls have evolved in chloroplasts (chlorophyll *b* in green algae and plants, and chlorophyll *c* in brown algae and in diatoms). A type of cyanobacterium (*Prochloron*) exists that also has chlorophyll *b*; but other evidence indicates that chloroplasts derive from more typical cyanobacteria and that other chlorophyll types have evolved independently in different chloroplast lineages. Typical cyanobacteria have a number of other accessory pigments: phycocyanin and phycoerythrin that allow for light absorption in the orange and green parts of the spectrum. These have been lost in most chloroplast lineages, but have been retained in the chloroplasts of, for example, red algae. Chloroplasts code for a somewhat larger share of their proteins than do mitochondria. This may indicate that chloroplasts represent a later development.

There is little doubt about the origin of mitochondria and chloroplasts as prokaryote endosymbionts. The question is whether there are other eukaryotic organelles that can be explained in this way. One such candidate is the peroxisome. It has a single membrane and it does not contain DNA. This does not exclude the possibility of endosymbiotic origin since the entire genome could have moved to the host nucleus. The question remains open.

The successful explanation for the origins of mitochondria and chloroplasts resulted in a number other similar suggestions for explaining other eubacterial traits. In particular, Lynn Margulis has tried to explain the eukaryotic flagellum (and the centriole) as representing symbiosis with spirochaetes. Spirochaetes represent a group of filamentous bacteria. Their motility pattern superficially resembles that of flagella. The inspiration for this idea is that there exists a flagellate species on to which symbiotic spirochaetes attach by one end to its surface. The motility of these ectosymbiotic bacteria makes the protist swim as if the spirochaetes were its own flagella. There are, however, several difficulties with this model for the origin of flagella. First of all the mechanism of spirochaete swimming is quite different from that of flagella. Spirochaetes have a normal bacterial flagellum (a rigid, rotating corkscrew); the flagellum is internal, mechanically corresponding to a rigid helix rotating inside a rubber tube. In eukaryotic flagella motility is due to sliding filaments. Attempts to demonstrate tubulin in spirochaetes have so far been unconvincing. Neither flagella nor centrioles contain DNA and a direct division of centrioles has never been demonstrated.

It is tempting to try to derive further eukaryotic traits from symbiotic relationships. So far, however, the only hard facts appear to be the origin of mitochondria and chloroplasts.

Models for evolution from symbionts to organelles

The term *symbiosis* is often used to describe an association of mutual benefit. One might think of ants that move aphids to fresh leaves and the ants in turn profit by feeding on the sugary excretions of the aphids. Biologists usually apply the term *mutualism* to describe such relationships that increase the fitness of both participants. The term symbiosis then means only that two types of organisms live in a physically close association, but the term does not imply who profits from co-habitation. Symbiosis thus also includes parasitism and various almost neutral associations.

In the case of endosymbiosis, that is, in which one type of cell (the symbiont) lives within another type of cell (the host) it is often meaningless to speak of mutualism. What has happened is that a microorganism somehow entered the future host cell, most likely through phagocytosis. For some reason the ingested cell got out of the food vacuole and thus evaded digestion. Under some circumstances the ingested cell could survive and reproduce within the cytoplasm of the host cell and its descendants. The relationship is stabilized if the presence of the symbiont somehow favours the host or at least that the association is initially neutral. At this point it is meaningless to ask whether the association is beneficial for the symbiont to be encased within host cells, because it no longer competes with its free-living relatives. In many ways one may consider the relation as one in which the host cell parasitizes the symbiont. It is not entirely unlike the way we keep chickens and cows; whether they 'benefit' from this relationship is a pretty meaningless question. A hereditary improvement in meat, egg, or milk production favours the farmer and is selected for. The farmer will also adopt procedures that optimize growth and reproduction of his animals. So this kind of relationships would also appear as a mutualistic system although, in fact, we eat the animals or otherwise exploit them.

If there is any advantage for the host cell and its descendants to harbour the symbiont, then they will eventually out-compete cells without symbionts. The relation has become obligatory and in principle a new unit for natural selection has appeared. In general, mutations in the symbiont that favour the host will be selected for because the particular host, and thus its symbionts, will become more numerous. Conversely, symbiont mutations that are unfavourable to the host will be selected against.

This latter statement is not, however, always true. The mutual competition among symbionts within a host cell might lead to the breakdown of the association. Thus a mutation that increases the growth rate of a symbiont cell could be favoured even though it might carry traits that are dysgenic to the host. In general, however, the symbionts within a host cell are clones that descend from a single symbiont cell within the host cell, and this minimizes this risk for evolutionary instability (see also Chapters 5 and 9). In sexual host species it is typical that symbionts (organelles deriving from symbionts) are exclusively inherited from one sex. Mitochondria, for example, are inherited maternally in animals. This arrangement minimizes the effect of intraspecific competition between the symbionts (or organelles) within a cell. Another stabilizing effect is the horizontal gene transfer from the symbiont to the host genome.

There are innumerable examples of types of endosymbiosis in which certain prokaryotes or small eukaryotes live inside the cells of other eukaryotes. In many cases the functional significance is unknown. In some cases it has been shown (e.g. through removal of prokaryotic symbionts using antibiotics) that the host has become dependent on the symbiont; the reason is probably often that the symbiont synthesizes a vitamin and the host cell has lost this ability. There are also many rather exotic examples of symbiosis. Some ciliates use bacteria as a sort of biological warfare against other competing ciliates. Some fish and cephalopods harbour luminescent bacteria in their light organs.

The type of symbiosis of interest in the context of explaining the origin of cell organelles is one of syntrophy. That is, that one component produces a metabolite that is used by another component that again somehow favours the first one. An interesting property of such interactions is that they can be shown to be instantly stable in the sense that the growth rates (generation times) of the host and its symbiont become identical. This is similar to pairs of syntrophic bacteria (see Chapter 7).

The development from symbiont to organelle is in a sense completed at the moment when the symbiont can no longer survive outside the host cell. This is likely to happen in unicellular hosts where the host is transferred vertically from generation to generation of host cells. In multicellular hosts this is a less probable event. This is because symbionts are most likely to reside in somatic cells with a built-in limited life span. In this case symbionts must at intervals leave host cells and survive outside until it has found new host cells. Symbionts are less likely to reside in cells of the germ line that are potentially immortal. Endosymbionts are common in many animals. An example is provided by photosynthetic algae that live in tissue cells of, for example, corals and some molluscs. But in these cases evolution to the status of organelle is less likely. However, there are some examples where symbionts are transferred via gonadal cells and with a potential to develop into organelles.

In unicellular hosts, evolution towards organelle status may take place quickly (in an evolutionary scale) in terms of loss of functions that are provided by the host cell. The analysis of many cases of endosymbiosis also show that over longer time-scales genes tend to be transferred from the symbiont to the host. It is conceivable how this may happen and why the process is unidirectional. There are usually many symbiont cells in each host cell. Once in a while a symbiont cell will succumb and lyse and its genetic material will float around in the cytoplasm of the host cell and may eventually become incorporated in the host genome. Once the corresponding proteins are synthesized by the host, loss of the corresponding genes in the symbiont is likely to be selectively favoured. Finally it is also possible that gene transfer to the host is selectively favoured because it minimizes inter-symbiont competition.

Examples of endosymbiotic cyanobacteria (so-called *cyanella*) are not common. A well-studied example is the flagellate *Cyanophora paradoxum*. But although the cyanellae look almost like free-living cyanobacteria with a reduced cell wall, genetic analysis has shown that a large share of their genome is already found in the nucleus of the host. Other cases have been less well studied. Eukaryotic photosynthetic endosymbionts, on the other hand, are very common. There are examples of almost all stages between symbionts that can still be grown outside of the host (e.g. the ciliate *Paramecium bursaria* that is filled with cells of the green alga *Chlorella*) (Figure 8.3; see also Plate 2) and strongly integrated and reduced symbionts that have basically attained a status as organelles (as in the above mentioned cryptomonad flagellates).

Some protozoa and some nudibranch gastropods have the ability to retain chloroplasts from their ingested food algae and to maintain them in a functional state over periods of days or weeks. These isolated chloroplasts deliver photosynthate (in the form of carbohydrates) to the 'host' cell. This represents a sort of evolutionary cul-de-sac in the sense that the chloroplasts will never regain their ability to divide in the absence of the genome in the algal cells. This phenomenon cannot lead to organelles.

Models for the origin of mitochondria are less frequent, but a couple of examples (Figures 8.5 and 8.6) may represent realistic analogies. The first example (Figure 8.5)

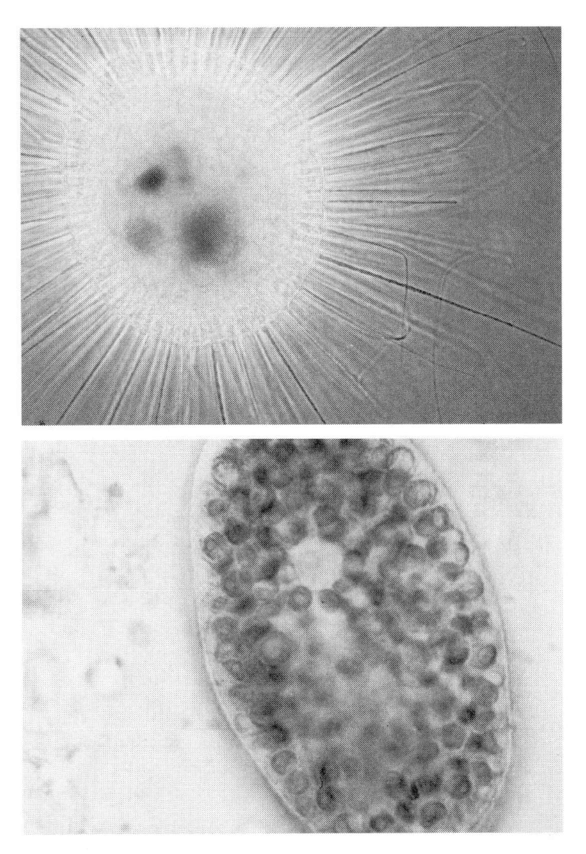

Fig. 8.3 (See also Plate 2) Eukaryotic organisms. (Top) The heliozoan *Actinosphaerium eichorni*, a common unicellular freshwater organism that may reach a size of about 1 mm. It is a predator feeding on ciliates and other protists; the brown areas are ingested prey. (Bottom) The ciliate *Paramecium bursaria* containing hundreds of endosymbiotic green algae (*Chlorella*). The ciliate measures about 100 μm and its symbionts measure about 5 μm. (Originals)

is the secondarily anaerobic ciliate *Strombidium purpureum*. It occurs in shallow marine sediments often together with sulphur bacteria. The purple colour of this creature is caused by masses of intracellular purple non-sulphur bacteria. In this way the ciliate is photosynthetic and it accumulates in near infrared light in accordance with the absorption spectrum of the symbionts. In the light, the ciliate prefers anoxia. The symbionts then perform photosynthetic activity based on the fermentative metabolites of the host (hydrogen, low molecular weight organic compounds). In the dark, the ciliate is attracted to places with a low oxygen tension and under these circumstances the bacteria can respire using O_2. The example is interesting in that symbiosis (for the second time) has converted an anaerobic eukaryote into an aerobe (at least in the dark). If the symbionts would lose their bacteriochlorophyll this would be a perfect analogy for our ideas about the origin of mitochondria.

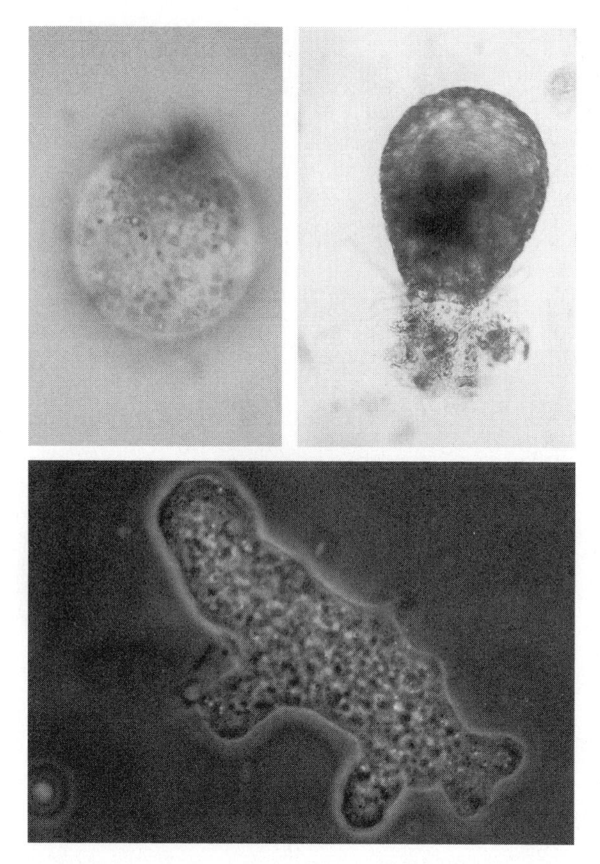

Fig. 8.4 (See also Plate 3) Eukaryotic organisms. (Top, left) The ciliate *Strombidium purpureum* containing endosymbiotic purple non-sulphur bacteria (see also text and Figure 8.5). (Top, right) The about 100 μm long testate amoeba *Difflugia* that covers itself with mineral particles; a couple of pseudopodia can be seen sticking out of the test below. (Bottom) A typical naked amoeba. (Originals)

The other example (Figure 8.6) is analogous to the origin of mitochondria in the sense that it involves endosymbiotic bacteria that utilize the metabolites of a fermentative host. The symbionts are not aerobes, but methanogenic bacteria. The phenomenon is known from a number of anaerobic ciliates, certain flagellates, and an amoeba-like organism. Genetic analysis of the hosts and their symbionts shows that this type of symbiosis has evolved independently on several occasions. The symbiosis is an example of hydrogen transfer syntrophy as shown in Figure 7.10. The fermentative metabolism of the host depends on the maintenance of a low H_2 tension within the cell. This is accomplished by the methanogens that oxidize H_2 with CO_2 to produce CH_4. Experimental removal of the methanogens leads to a decrease in H_2 production of the ciliates and to a decrease in their growth rate.

Altogether the examples presented in this section show that organelle evolution from endosymbiosis is not something that took place only 1.5–2 billion years ago, but

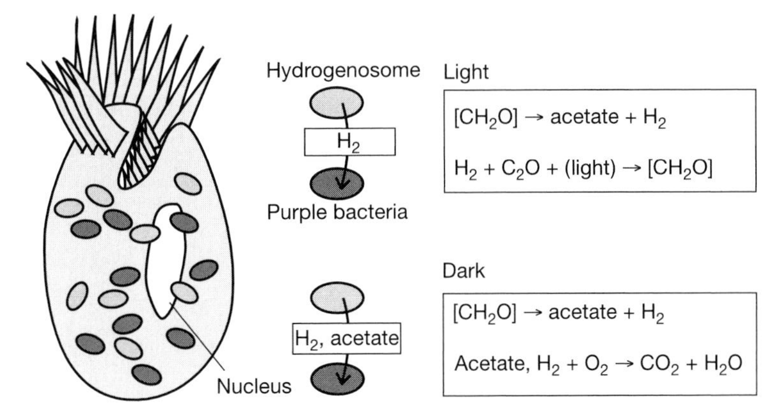

Fig. 8.5 The anaerobic ciliate *Strombidium purpureum* harbours symbiotic purple non-sulphur bacteria. In the light the bacteria use the fermentative metabolites of the host for photosynthesis. In the dark the bacteria carry out aerobic respiration and so oxidize the host metabolites. This is perhaps the best-known analogy to the origin of mitochondria.

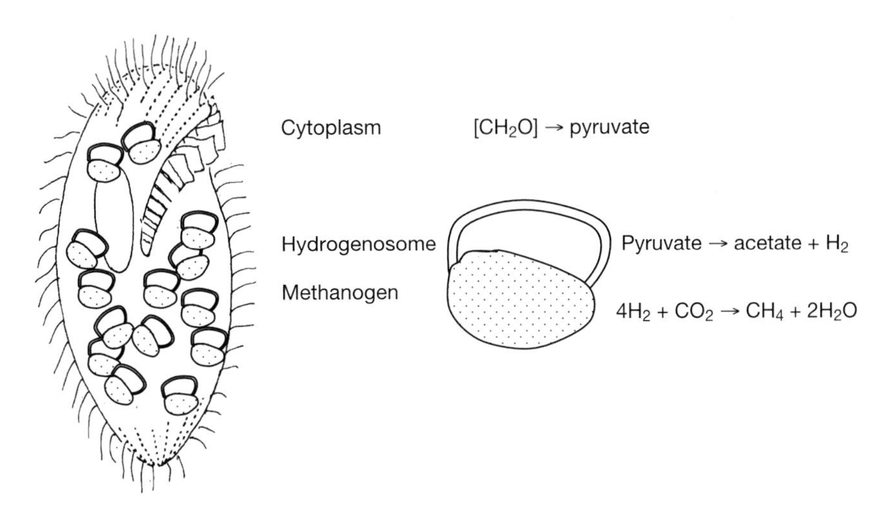

Fig. 8.6 The anaerobic ciliate *Metopus contortus* harbours symbiotic methanogenic bacteria. These use H_2 produced by the fermentative metabolism of the host; at the same time the bacteria secure a low internal H_2 tension, thus making host metabolism more effective.

that evolution through chimera formation between eukaryotes and prokaryotes still happens. The origin of chloroplasts and of mitochondria had an enormous effect on the biosphere. In the extant biosphere most of the light-driven fixation of CO_2 is due to chloroplasts in eukaryotes (plants on land and algae in the sea) and a very large share of respiratory O_2 reduction is due to the descendants of purple bacteria that became mitochondria.

Chapter 9

Multicellular organisms—origins as cell colonies

The multicellular organisms represent the next major transition in the evolution of life (after the origin of the first prokaryotic cells and the origin of the eukaryotes). Three groups: fungi, plants, and animals have evolved independently, but they remain relatively close to each other on the eukaryote phylogenetic tree (Chapter 12, Figure 12.8). The fungi descend from a group of fungi-like protists (the chytrids), the plants from green algae, and the animals from a group of flagellates (choanoflagellates). Green algae, red algae, and brown algae are often (with good reason) included among the multicellular organisms; they also form large cell colonies that include different functional types of cells and many of these algae grow to a quite large size. Finally, many representatives of various protist groups (and even some bacteria) produce cell colonies that share some features with true multicellular organisms.

The multicellular organisms appeared relatively late in the history of Earth. Fossils of what is believed to have been multicellular algae are known from about 600 million-year-old deposits. The oldest known remains of animals are about 590 million years old. Molecular evidence, however, suggests that the major groups of invertebrate animals had already diverged prior to this time. The first animals may have been small (millimetre sized) and soft bodied like many extant forms such as rotifers, nematodes, turbellarians—and this probably explains the apparent absence of earlier fossil evidence. Vascular plants originated even later (during the Silurian and Devonian times) in connection with the appearance of life on land, and flowering plants go only some 120 million years back in time. The most important role of fungi is to decompose plant tissue in terrestrial soils; their great diversification is probably also relatively recent in geological terms. This diversification was probably also an effect of the widespread symbiotic relationships with vascular plants in the form of mycorrhiza formation.

Here it is not the intention to account for the many complex properties of plants and animals, their evolution, or their diversity. We will consider only their basic properties, how they evolved, and (in Chapter 14) why they occurred so late in geological time.

The tendency to form cell colonies among various unicellular eukaryotes is widespread. Usually they result from incomplete cell divisions, or less frequently, from single cells that fuse together during certain stages of their life cycle. The adaptive significance of colony formation varies. The larger colonies may enjoy a lower risk of being eaten than single cells and sometimes it can be shown that cell colonies have a more effective food uptake. In some cases multicellularity serves for producing sporangia that can spread propagules more effectively. Altogether, selection for larger body size has been a driving force in the evolution of multicellularity.

The specialization of different cells to carry out different tasks is a decisive feature. One such fundamental specialization, also known from some protist cell colonies, is that only some of the cells are reproductive (capable of division or of sexual processes). This applies to, for example, colonies formed by some peritrich ciliates (Figure 9.1). While all cells are capable of dividing to produce new members of the individual colonies, only certain cells are capable of producing swarmers that can establish new colonies. The individual colonies have a limited life span.

Multicellularity implied built-in death. Unicellular organisms, of course, die all the time for a variety of reasons: being eaten, lack of resources, etc. But a unicellular organism is potentially immortal, meaning that in principle it could go on growing and dividing forever. This is not so for us (animals, and to some extent plants and fungi).

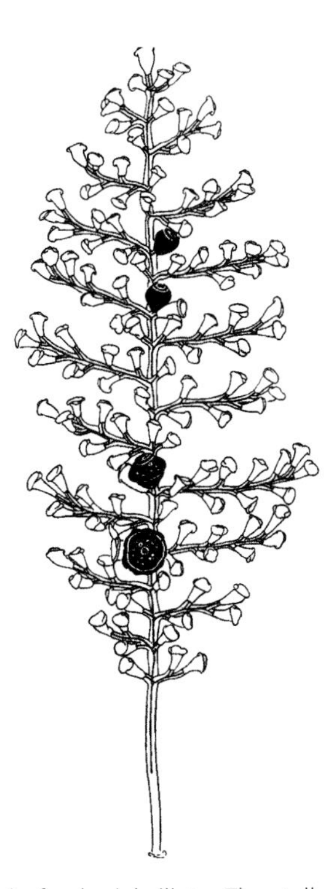

Fig. 9.1 *Zoothamnium* is a kind of colonial ciliate. The stalks of individual cells do not detach from one another during cell division and so branched, tree-like colonies form from the founder individual. All cells are capable of division, but only special cells (coloured black in the figure) are capable of producing free-swimming cells that can found new colonies. The individual colonies have a limited life span. From K. G. Grell. (1968). *Protozoologie*. Spinger–Verlag, Berlin.

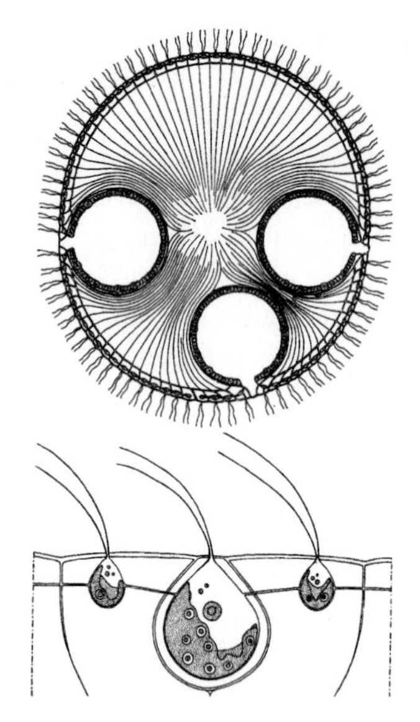

Mother volvox sacrifices herself that sex may be introduced into the animal world; her sexually produced offspring weep about her bier.

Fig. 9.2 The green alga *Volvox* forms spherical colonies. Gametes are released inside the colony where fertilization and formation of new colonies also take place. These new colonies can escape only after disintegration of the mother colony. From K. G. Grell. (1968). *Protozoologie*. Spinger–Verlag, Berlin. (Right) Built-in death enters the living world. From R. W. Hegner. (1938). *Big fleas have little fleas or Who's who among the protozoa*. Williams and Wilkins, Baltimore. Reprinted by Dover, New York, 1968.

An example of built-in death is shown in the green alga *Volvox* (Figure 9.2). This organism, common in the plankton of lakes and ponds, consists of spherical colonies in which individual cells are held together by mucous threads. New colonies are formed sexually from gametes that are released inside the colonies. After fertilization, the zygotes form new colonies within the old one, and the new colonies only escape when the mother colony succumbs.

It is therefore a characteristic of multicellularity that some cells sacrifice themselves in order to increase colony fitness. The potentially immortal part of multicellular organisms is the *germ line*, that is, the sequence of reproductive *germ cells* from generation to generation. For each generation some of the descendants of the germ cells form a colony consisting of *somatic cells* (a whole animal or plant) as shown in Figure 9.3. The Darwinian fitness of the colony determines the success of the germ cell in producing new colonies. The multicellular colony is the unit of natural selection, but it is discarded for each generation. This description applies to *Volvox* and largely also to animals, plants, and fungi. Some invertebrate animals, however, are capable of budding from somatic tissue, and these may perhaps also, in principle be immortal. This also applies to many plants and fungi. Furthermore, most plant cells retain the ability to develop gonadal cells and all other cell types. In animals, the future function of cells is often determined

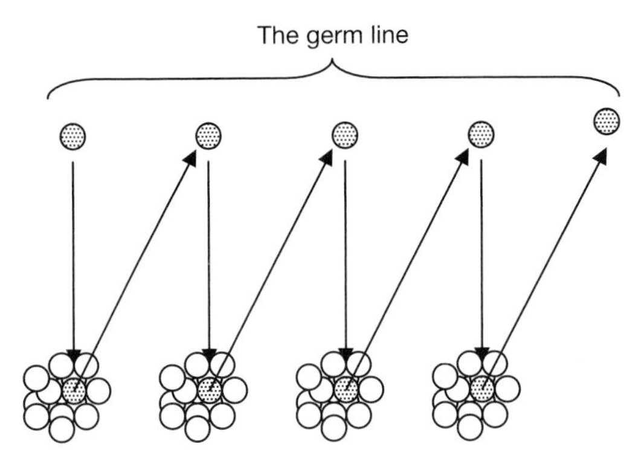

Fig. 9.3 The principle of multicellularity: only the cells of the germ line survive from generation to generation. For each generation a cell colony is formed, but it is discarded. The figure does not take sexual processes into account.

early in development and somatic cells are often capable of only a restricted number of cell divisions. In that sense one could say that animals are 'more multicellular' than are plants—also because the number of different specialized cells is much higher in animals than in plants.

The reason why such self-sacrificing cells could evolve is that the entire multicellular organism constitutes a clone, that is, they are genetically identical because they all descend from a single cell through mitotic divisions. Genes that result in the production of a colony of mortal cells will be promoted as long as this favours the fitness of the immortal germ cell through the production of a successful clonal colony. Once such a system was established the way was open for further specialization of somatic cells to carry out particular tasks.

It can be asked whether or not there is some fundamental reason why multicellular organisms have a limited life span, or in other words: why are we not immortal? It has been pointed out that the accumulation of somatic mutations determines the maximum life span. This is certainly a part of the ageing mechanism in animals. But then again, many unicellular sexless protists are in principle immortal in the absence of meiosis and recombination (see the next section) that would be required for DNA repair.

The fundamental answer is that eternal life is not adaptive for multicellular organisms. Consider an annual plant. A seed from such a plant is assumed to produce B seeds after one year and then it dies. That is, the population growth is B per generation (per year) and we consider B as a measure of fitness. Now suppose that one plant through some mutation acquires immortality so that it survives and produces B seeds every year forever. It can be seen that the gain in fitness equals that obtained by remaining an annual, but producing $B + 1$ seeds. And so the fitness gain from becoming immortal is rather trivial if B is a reasonably large number. An evolutionary change that would allow the plant to produce a few more seeds would appear more likely than a mechanism that allows for immortality.

Various factors may modify this result. If survival from seed to mature plant is poorer than survival of mature plants from one year to the following, then the fitness gain from becoming a perennial is higher. Also, if the time for maturation is longer than the interval between flowerings then increasing life span is also selected for. Trees must grow for many years before they produce seed, but from then on they do so annually and trees live for many years. Herbs in which maturation time corresponds to the time interval of seed production are frequently annuals. Brood protection, common in many animal groups, may also select for somewhat longer life span even beyond the fertile age classes.

Under all circumstances, fitness gain from an increase in life span decreases with increasing age. A phenomenon referred to as *genetic trade-off* is also relevant in this context. The term means that alleles that increase one fitness component simultaneously decrease another fitness component. Thus it is possible to select for increased longevity in fruit fly populations, but the resulting strains are less fertile. In nature these would then be less fit, so there is no selection for a longer life span. It is also realistic to assume that hereditable traits that provide greater resistance against old-age diseases (such as cancer) are coupled to an increased susceptibility to, for example, infectious diseases at a younger age. In such cases increased potential life span would have a negative net effect on fitness. When fitness increase from increasing life span becomes more marginal then the role of mutational load also increases (cf. discussion of the RNA world in Chapter 5).

There are different grades of multicellularity among animals. The sponges still show clear traits of a cell colony with a limited number of different kinds of cells. All other animals have a precise body plan and a large number of cell types, the function of which is determined early in development. New discoveries have shown that certain sets of tightly linked genes ('hox-genes' that regulate the transcription of other genes, so-called transcription factors) represent the basis for the body plan. These hox-genes are found in all animals except sponges and to some extent coelenterates. The basic body plan is principally the same for all animals from nematodes to humans, only with variations on a common theme. This basis was established before the major groups of animals diverged.

The duplication of units and subsequent specialization is a phenomenon that we have already considered at a lower organizational level, that is gene duplication and formation of cell colonies. The duplication or multiplication of certain parts of the body has been a frequent event during evolution. Examples include annelid worms and arthropods. In vertebrates the system of gill arches also exemplifies this. Later, the anterior-most gill arches became the jaws of primitive fish, and the following ones eventually became the auditory ossicles of mammals.

Colonial animals represent a kind of repetition of the development of cell colonies to multicellular organisms. In particular, coelenterates (e.g. corals) and tunicates form such colonies. Siphonophores are a kind of planktonic coelenterate; they are colonial and the different individuals have different functions. These include reproductive individuals (that are almost reduced to gonads), individuals that serve to catch food, and individuals that serve for motility (Figure 9.4). Again we see that individual units sacrifice themselves (in terms of being non-reproductive) for the

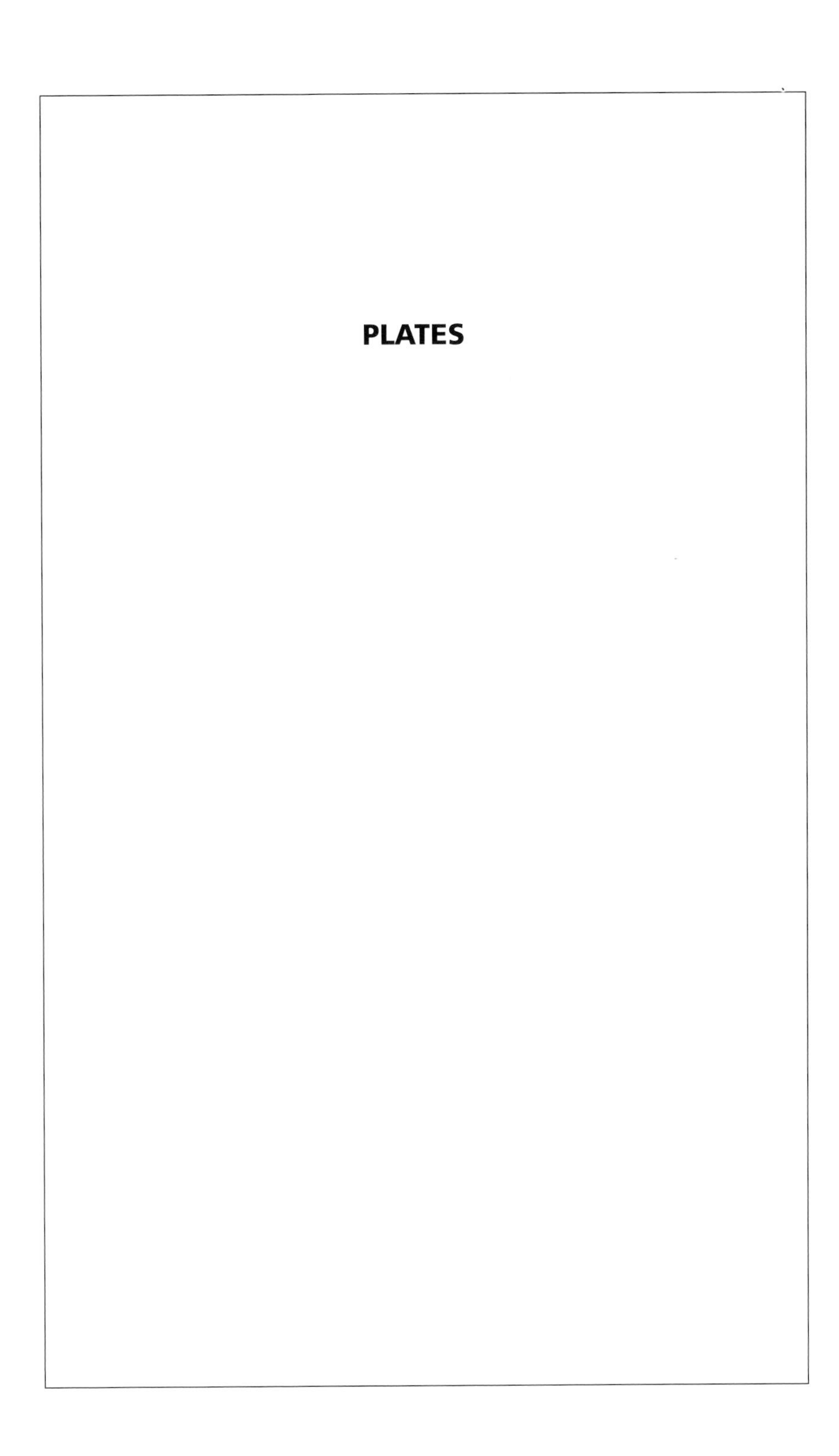

PLATES

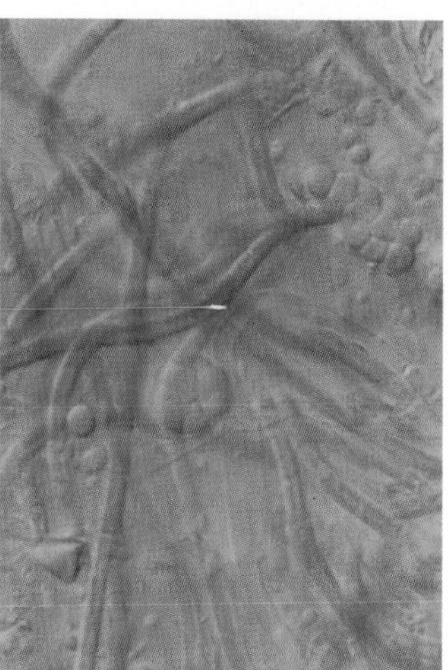

Plate 1 (Figure 4.6) (Top, left) Cultures of purple and green sulphur bacteria. (Top, right) The filamentous, colony-forming cyanobacterium *Pseudanabaena* and some other prokaryotic microorganisms. (Bottom) Microorganisms are not always invisible to the naked eye: the figure shows a shallow brackish bay (north of Copenhagen) totally covered by purple sulphur bacteria. (Originals)

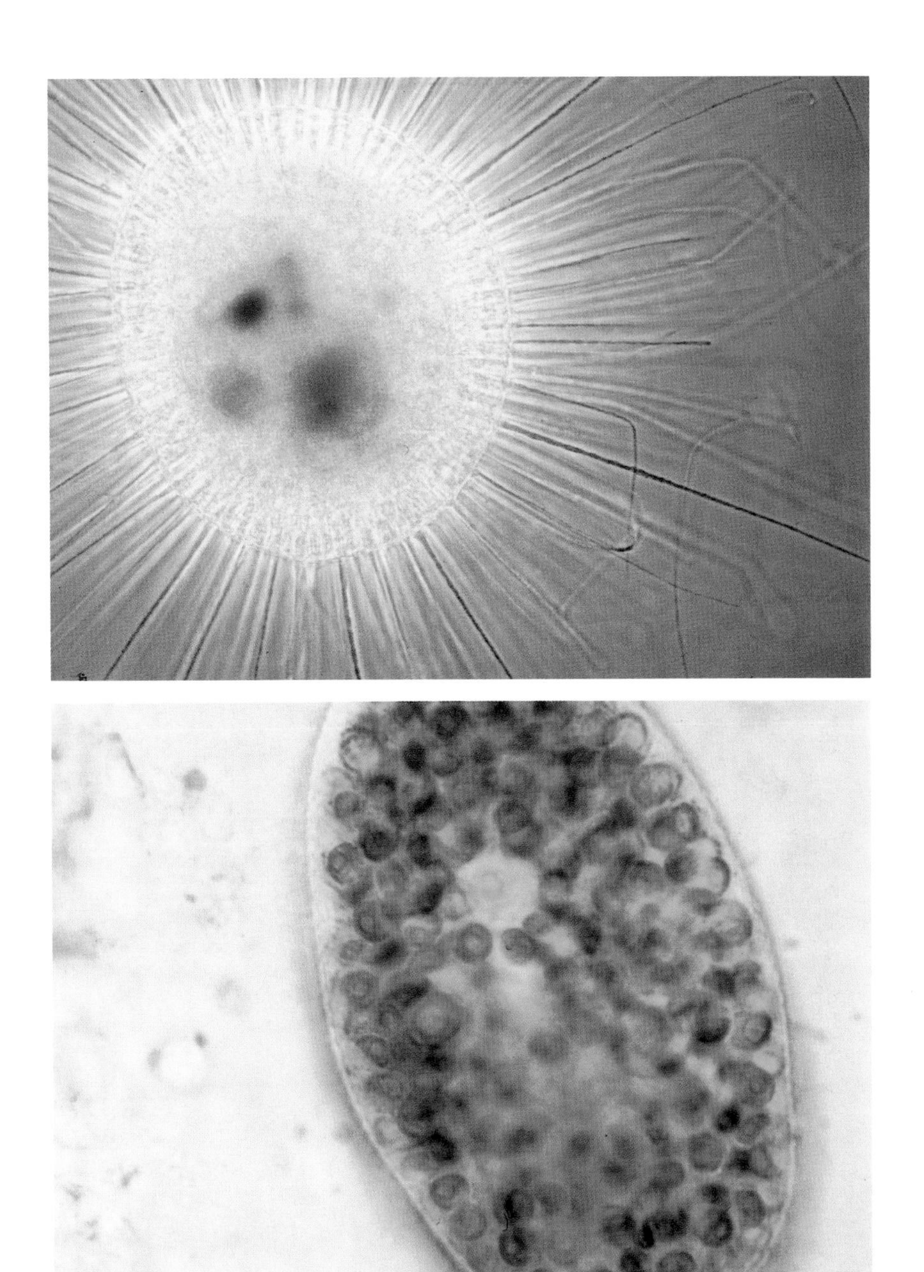

Plate 2 (Figure 8.3) Eukaryotic organisms. (Top) The heliozoan *Actinosphaerium eichorni*, a common unicellular freshwater organism that may reach a size of about 1 mm. It is a predator feeding on ciliates and other protists; the brown areas are ingested prey. (Bottom) The ciliate *Paramecium bursaria* containing hundreds of endosymbiotic green algae (*Chlorella*). The ciliate measures about 100 μm and its symbionts measure about 5 μm. (Originals)

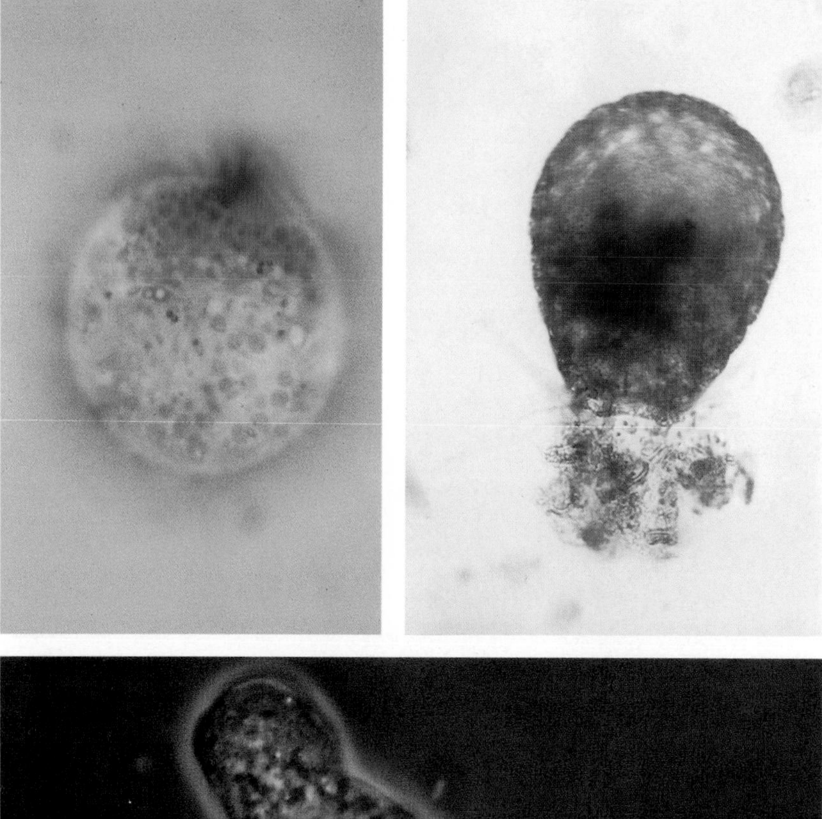

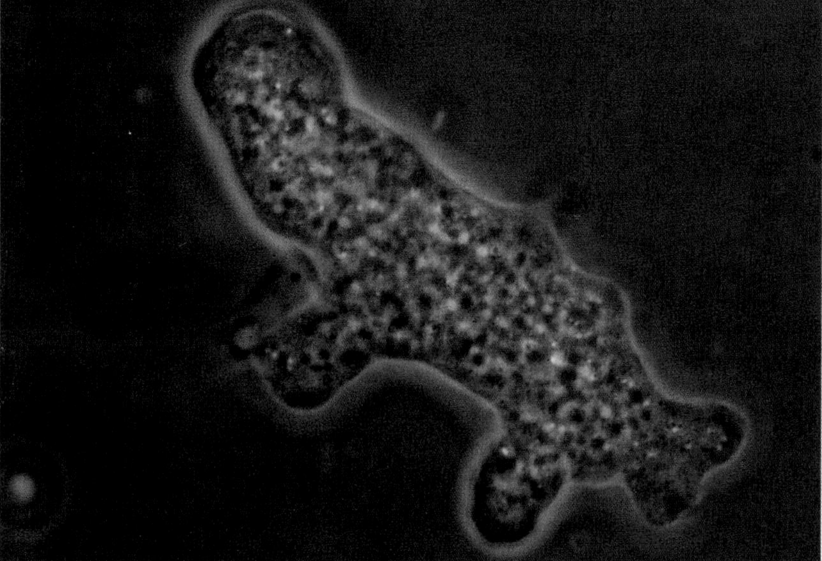

Plate 3 (Figure 8.4) Eukaryotic organisms. (Top, left) The ciliate *Strombidium purpureum* containing endosymbiotic purple non-sulphur bacteria (see also text and Figure 8.5). (Top, right) The about 100 μm long testate amoeba *Difflugia* that covers itself with mineral particles; a couple of pseudopodia can be seen sticking out of the test below. (Bottom) A typical naked amoeba. (Originals)

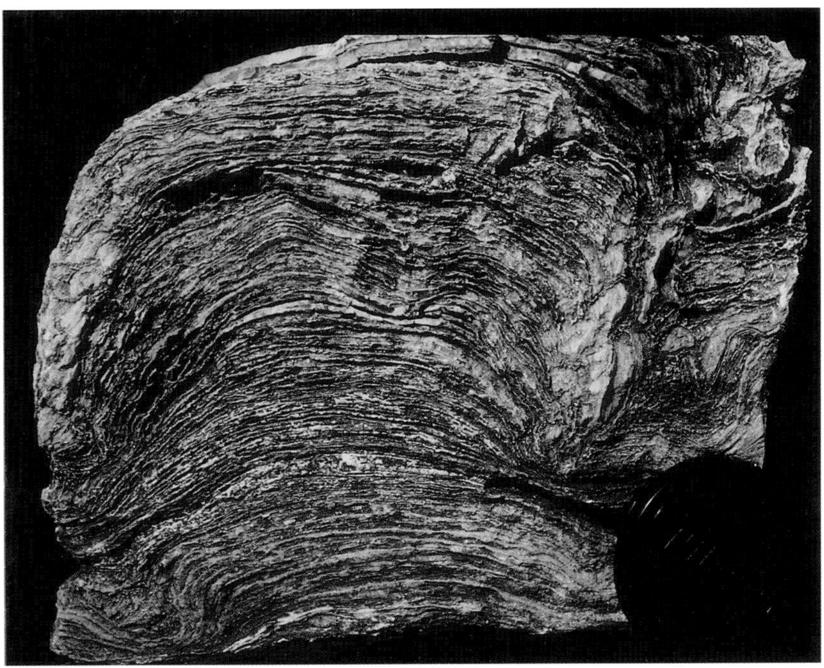

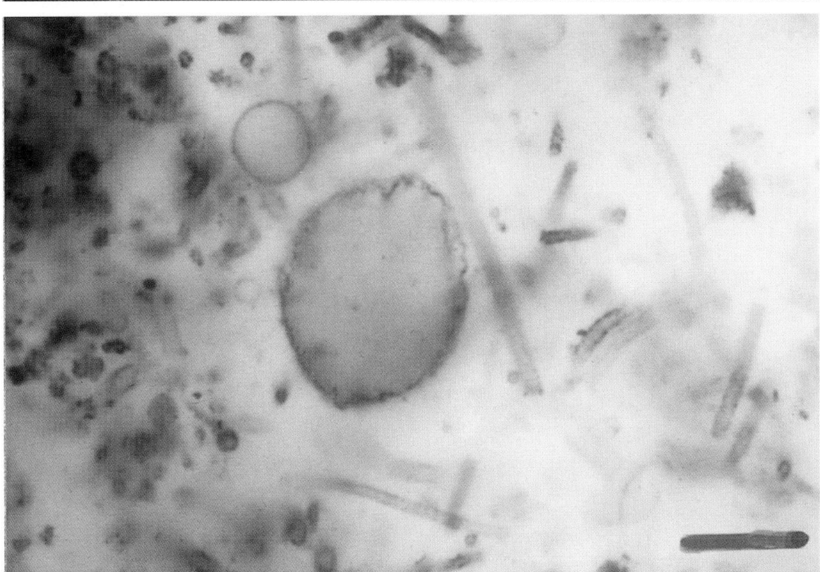

Plate 4 (Figure 13.3) (Top) A 2.2 billion-year-old stromatolitic rock from Belfast, Transwaal (South Africa). (Bottom) Thin section of an approximately two billion-year-old silicified Gunflint Formation, Canada, showing fossils of cyanobacterial filaments and of other microbes. (S. M. Awramik)

Plate 5 (Figure 13.4) (Top) Cyanobacterial mats in a hot spring in Yellowstone Park. (Bottom) The hypersaline lake 'Solar Pond' in the Sinai Desert. The bottom is covered by a stromatolitic cyanobacterial mat roughly one metre thick, the surface of which is responsible for the yellow–green colour. Perhaps the picture provides an impression of shallow water habitats through three billion years of the Earth's history: the land is barren and apparently lifeless while the bottom of shallow water habitats is covered by stromatolitic microbial mats. (M. Kühl)

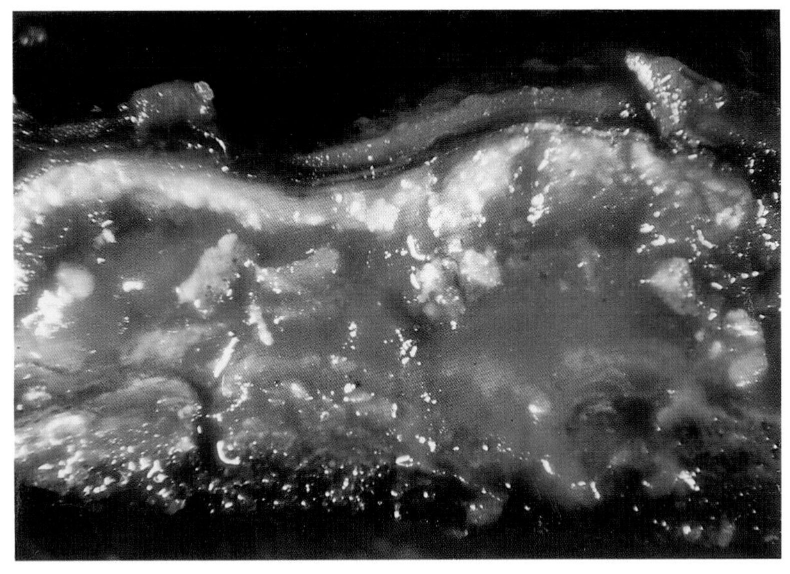

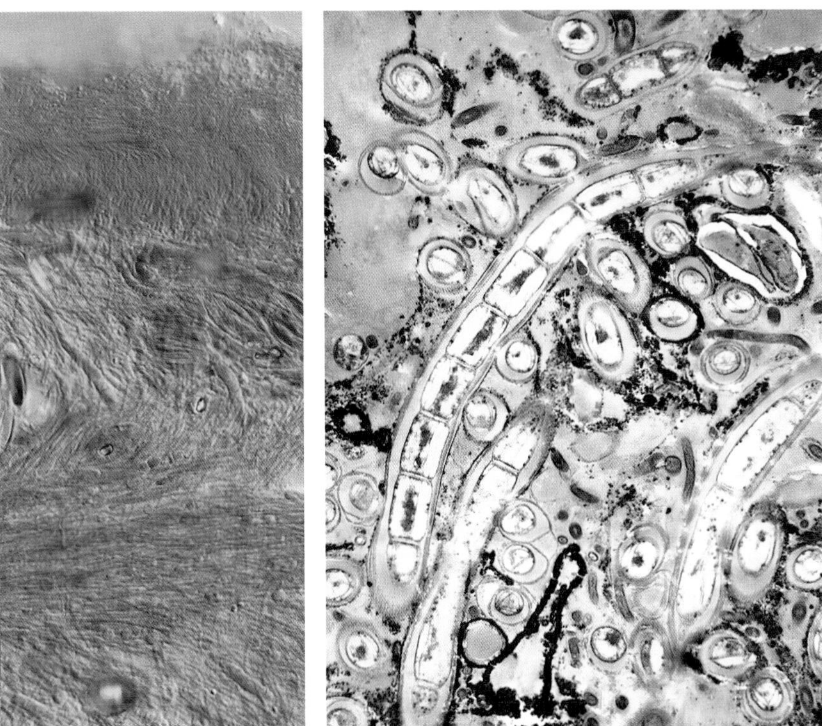

Plate 6 (Figure 13.5) A section through an artificial stromatolitic cyanobacterial mat grown in the laboratory in the absence of animals. It is about three years old and 7 mm thick. Yellowish and dark green layers with cyanobacteria are seen in the upper 1 mm above a layer with carbonate precipitation. The red, brownish, and greenish colours deeper in the mat are due to different types of bacteria with anoxygenic photosynthesis. (Bottom, left) A freeze-section of the upper 2 mm of the mat with filaments of the cyanobacteria *Calothrix*, *Pseudanabaena*, and *Phormidium*. (Bottom, right) Electron micrograph about 1 mm beneath the mat surface including *Pseudanabaena* filaments and various other bacteria. (Originals)

Plate 7 (Figure 13.7) Banded iron formation in Western Australia. (D. E. Canfield)

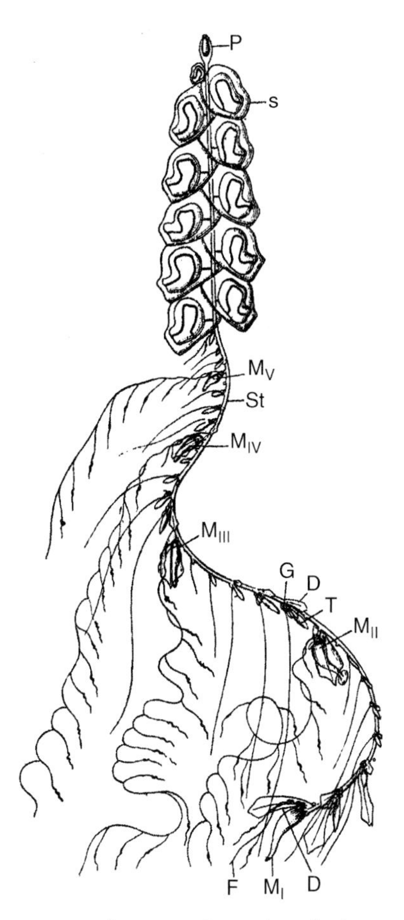

Fig. 9.4 The siphonophoran *Cupulita* occurs in marine plankton. It is a colony consisting of several types of individuals that are specialized to perform particular tasks. S, individuals that serve only for motility; M, feeding individuals with sticky threads for catching prey; P, pneumatophor, a gas-filled individual that controls buoyancy. Only the gonad-containing individuals (G) are reproductive; all other individuals sacrifice themselves for the benefit of the fitness of the colony as a whole.

well-being of the entire colony. And again, this altruism is conditioned on the fact that the colony is clonal and results from vegetative reproduction (budding). Colonial animals provide yet another example showing that the complexity of organisms during evolution has been attained mainly through the integration of simpler modules.

Chapter 10

Sex, species concepts, and evolution

Sex

Although horizontal gene transfer is known from bacteria it is not possible to speak about bacterial sex in the sense that it applies to eukaryotes. Many groups of unicellular protists also seem to be sexless. This applies, for example, to naked amoebae: the classical textbook amoeba *Amoeba proteus* has been grown in laboratories for almost a century, but no sexual phenomenon has ever been recorded. Totally sexless forms also occur among many groups in which other representatives display some sort of sexual phenomena. Many groups of unicellular eukaryotes display different components of sexual processes and in other cases protozoan sex resembles that of animals. Reproduction is not coupled to sex in unicellular eukaryotes as is normally the case in animals and plants. In protists reproduction results from mitotic divisions whereas sex, if present, occurs sporadically in connection with specific environmental conditions or as an obligatory part of a more complicated life cycle. This diversity may, perhaps, make it easier to understand the origin of sex.

For a start we can observe that most eukaryotes are *diploid* during parts of their life cycle. This means that the nucleus contains a double set of all the chromosomes. In this way detrimental mutations in one of the chromosomes are less likely to affect the phenotype and we may assume that this constituted the original adaptive significance of being diploid.

During *meiosis* a diploid cell produces four *haploid* cells (haploid means that there is only one set of chromosomes). In multicellular organisms and in some protists, three of the four haploid cells degenerate so that only one is eventually left. A typical meiosis is shown in Figure 10.1. Briefly, meiosis is initiated by pairing of homologous chromosomes in a diploid cell. This is followed by recombination so that parts of the two homologous chromosomes change place. This is followed by two successive cell divisions without chromosome doublings resulting in four haploid cells. In many eukaryotes the haploids undergo many successive mitotic divisions (Figure 10.2). In animals the haploid generation is represented only by egg and sperm cells. But in many primitive vascular plants (mosses, ferns) there is also a complete haploid generation represented by special phenotypes; remains of this also occur in flowering plants. In all cases, at some point the haploid cells (or a multicellular haploid generation) produce gametes. These fuse with another gamete to produce a diploid *zygote*—the first cell in a diploid generation. In many unicellular eukaryotes there are variations on this theme; for example, recombination may be absent. Some forms have special mechanisms that serve to avoid self-fertilization. These mechanisms are

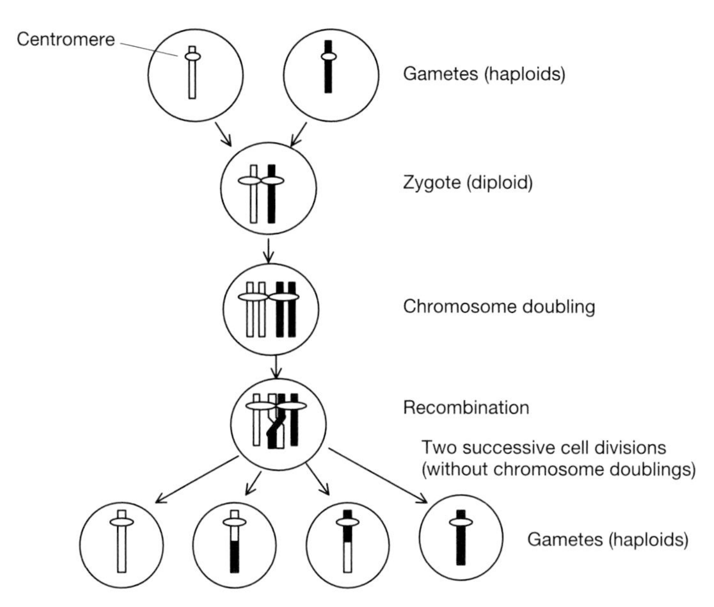

Fig. 10.1 Life cycle for a sexual organism with alternation between a diploid and a haploid generation. For further explanation see text.

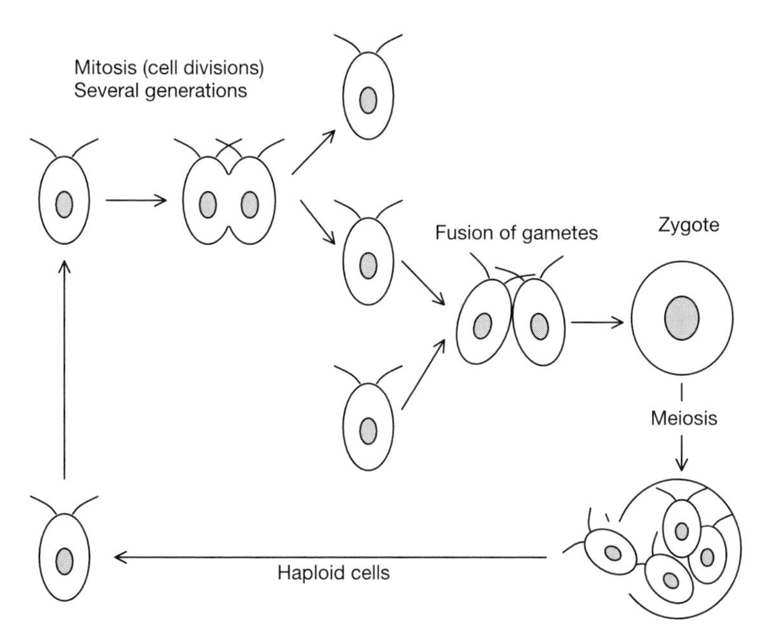

Fig. 10.2 Life cycle of the flagellate *Chlamydomonas*. Haploid cells can divide by mitosis for many generations. At some point these transform into gametes that fuse to form zygotes. The zygote immediately undergoes meiosis to produce four haploid cells.

analogous to gender in animals, but in certain groups there are complex systems with more than two compatibility groups, or more than two genders. Such mechanisms guarantee out-breeding. In other groups, however, haploid cells deriving from a meiosis immediately fuse to form a zygote, that is, obligatory inbreeding. In other forms recombination takes place, but the last meiotic division can be absent so that a diploid cell is the end-product of meiosis.

This diversity, in particular among the unicellular eukaryotes, suggests that sex represents several different mechanisms that had, or have, different types of adaptive significance. Recombination leads to formation of new genotypes, but it also brings DNA repair, and this last effect may have been the original role of recombination.

It has been pointed out that without sex and recombination it would be impossible to sort detrimental mutations from useful ones that reside in the same genome. Also, the diploid situation hides detrimental recessive mutations, and in the absence of sex and recombination such genes would eventually accumulate in the population. As an explanation for the origin of sex these arguments are difficult because they imply the future evolution of the populations—that is a sort of group selection, but not individual fitness. A way out of this dilemma is to assume that asexual populations originate at intervals and thrive for some time, but that they eventually succumb to the above mentioned problems. Examples of this are not rare among plants and some invertebrate groups. Apomictic dandelions, for example, are common, but these lineages are believed to become extinct after some time while new apomictic strains evolve from sexual plants. (Apomixis means that seeds develop without fertilization.)

In many groups a sexual life cycle is fixed for quite different reasons. Certain cell organelles are transferred only by gametes from one of the sexes and zygote formation is therefore necessary. Thus sperm cell provide the centriole in mammals and in conifers the chloroplast is transferred with pollen only.

A model that might explain the maintenance of sex is the 'lottery model'. A tree, for example, produces thousands of seeds. Most of these are unsuccessful, but some will end up in patches where they may possibly grow. Each of the patches represents particular properties with respect to humidity, light, soil, etc. Now if the tree would be asexual then all the seeds would be genetically identical. But through recombination and out-breeding, no two seeds will be genetically identical. The probability that a given seed will be particularly fit in some particular patch would therefore be maximized. A somewhat similar consideration is the following: if an individual performs poorly due to deteriorating external condition, then a gene might increase its fitness by moving into another genotype that may perform better under the given circumstances and so induce sex. Actually in many protozoa sex is induced when the environment deteriorates, for example, due to lack of food. Genes that induce sex thus increase their chance of getting into another phenotype that might prove more robust under the prevailing conditions.

What is a species?

Darwin's most influential book is entitled *Origin of species*, although, in fact, it is mainly concerned with the evolution of lineages. Nevertheless, the species and the species concept are generally assumed to hold central positions in evolutionary theory. This view is strongly influenced by our understanding of the evolution of sexual, out-breeding organisms and especially of animals and plants. Insofar as prokaryotes and many unicellular eukaryotes do not show sexual phenomena, it is clear that the early evolution of life in certain respects deviates from the more familiar ideas about speciation—notwithstanding the existence of horizontal gene transfer. With this background it will be appropriate to discuss the species concept with special reference to microbial evolution.

Taxonomists operate with a hierarchy of groups: phyla, classes, orders, families, genera, and species (Figure 10.3). This system serves two purposes. It is a way in which a huge amount of information can be ordered and compressed. Various traits correlate within particular groups as already observed by Aristotle. Ruminants, for example, have horns or antlers, hoofs, a rumen, and they lack incisors in their upper jaw. So when it has been established that a given animal is a ruminant, then it is also established that it has a whole set of properties. The biological classification system thus holds a vast amount of information and as such it is invaluable. The other goal of biological classification is to describe how different groups of organisms are related. Organisms are thus classified in groups (a genus, for example) so that all members of the group have a common ancestor. The classificatory system thus includes our current understanding of the mutual relatedness of organisms in a genealogical sense. New insight with respect to phylogeny may lead to changes in this system. It is important to emphasize that the biological classificatory system is not an arbitrary construction. The taxonomist does not invent a genus or a family: he discovers them!

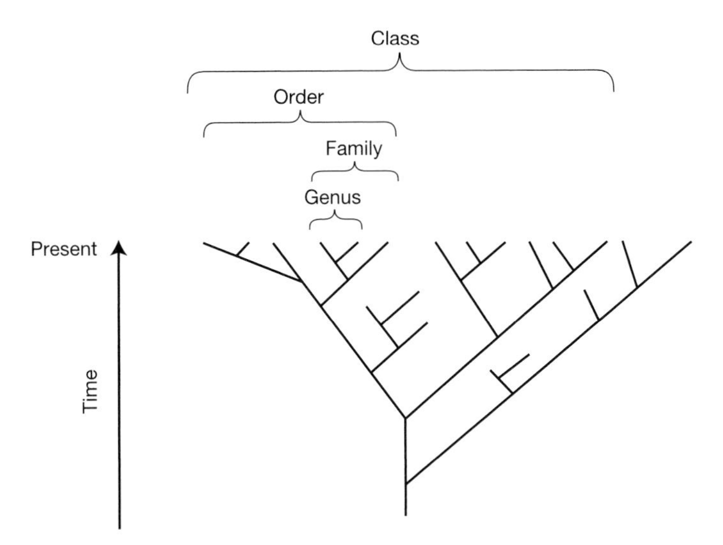

Fig. 10.3 The hierarchy of taxonomic categories. The system can be refined by introducing, for example, superfamilies and subfamilies. All members of a category (e.g. a family or a genus) share a set of common characteristics and are assumed to have a common ancestor (that is, every category should be monophyletic).

Among classificatory categories the species seem to have a special status. Prior to the acceptance of biological transformation it could be argued that species resulted from divine creation whereas higher taxonomic categories simply reflected abstract human book-keeping. Organisms always seem to fall into discrete groups, at least within a certain restricted geographical region. An example could be blackbirds, song thrushes, fieldfares, etc. No two blackbirds are quite alike, but all blackbirds are much more similar to one another than to any other kind of thrush. The species concept therefore seems self-evident and natural and it was established long before the theory of evolution.

In reality things are far from being easy and there are many definitions of what a species is. The *phenetic species concept* (or *morphospecies* when based only on morphological traits as is customary for macroscopic organisms) is based solely on a set of phenotypic properties that are shared by all individuals within a species. More recently it has also been possible to classify organisms based directly on the nucleotide sequences of genes (Chapter 12). In both cases, these approaches are generally applicable as a way to define the lowest level of classification, but the species concept has no special status relative to other taxonomic categories. The *biological species concept* plays an important role in the theory of evolution. But as discussed below it has meaning only for sexual out-breeders, and even in this case it often throws up difficulties. Sexual out-breeders are found especially among the more advanced eukaryote lineages. Most evolution, at least all early evolution, and branching of lineages occurred in a sexless world where the biological species concept does not apply.

According to the biological species concept, a species represents a population of interbreeding individuals. Such a population represents a common gene pool and it is at the same time genetically isolated from populations belonging to other species. This in itself explains why individuals within a species are rather similar. Speciation requires genetic isolation between two parts of a previously interbreeding population. The most important mechanism for this is probably geographical isolation. This could be brought about through geological or climatic events, or perhaps because a few individuals make it to an island or some other sort of isolated place. Over time the two now isolated populations will diverge due to genetic drift or to local selection pressure. If some event brings the two populations into contact again, different things may happen. The divergence may be so small that the populations merge. It is also possible that genetic divergence had become sufficiently great so that the meiosis of hybrids is affected during recombination, making them less fertile or sterile. Even greater divergence may mean that hybrid zygotes cannot develop. In these cases selection will favour mechanisms that prevents hybridization. Such mechanisms include temporal displacement of breeding periods, differential song patterns or plumage coloration in birds, etc. Speciation will now be a reality. Later selection may lead to differential resource or habitat utilization.

There are other ways in which populations can be separated into reproductively isolated subpopulations. In species that are also capable of vegetative reproduction, instant speciation may occur as a result of chromosome mutations such as chromosome doublings and inversions, or as a result of hybridization. When the mutant's vegetative offspring have become sufficiently numerous, they can then interbreed sexually and a new species can arise. This is an especially common speciation mechanism in plants.

Speciation is not an adaptive process *per se*. It is only a consequence of sexual reproduction and out-breeding and from the fact that meiosis and zygote development is ineffective or impossible if two genomes are too different.

The biological species concept includes several difficulties. Many 'species' form subpopulations *A*, *B*, and *C* where it can be shown experimentally that, for example, *A* and *B* and *B* and *C* can have fertile offspring together, but not *A* and *C*. Some species forms clines. They have a wide distribution, but change continuously from one end of the distribution range to the other. In such cases inter-fertility is possible within limited distances along the cline, but not between individuals from its extremes.

Finally it should be emphasized that there is no close correlation between genetic distance, phenotypic differences, and the ability to produce fertile offspring. Sibling species may be almost impossible to distinguish, but they are inter-sterile. Such complexes of sibling species are common, for example, among fruit flies; these are caused by chromosome inversions. In other cases, rather unrelated 'species' easily hybridize—this is especially common among plants. Finally, even among invertebrate animals and especially among plants and fungi there are many asexual, parthenogenetic, or self-fertilizing forms, and here the biological species concept does not apply. It is still a somewhat open question to what extent such organisms fall into discrete phenotypes, or whether they may represent continua of phenotypes.

The expectation of finding discrete species applies, of course, only to a specific point in time. Over time, lineages may change gradually, eventually representing a 'new species', but the exact moment when this 'transition' takes place probably cannot be known.

All this 'mess' can be predicted from evolutionary theory. In addition to this, the biological species concept is often not operational because there is no possibility to determine whether two organisms can produce fertile or viable offspring. This, for example, is the case for the zoologist working with collections of dried insects or formaldehyde-pickled fish. In practice, species of the great majority of animals and plants are defined from phenotypic traits and, in particular, external morphological traits that are visible in preserved specimens. In most cases the phenetic and the biological species concepts probably coincide. But the phenetic species concept does not have any special status among other taxonomic categories.

Species concepts for microbes—evolution without sex

Prokaryotes do not have sex and the biological species concept does not apply. If bacteria were completely clonal (that is, that recombination or horizontal gene transfer never takes place), then one might say that every beginning of a new clone—that is every cell division—represents a speciation event because the two new cells can each initiate two lineages that are in principle forever genetically isolated from one another. Each of them may over time accumulate predominantly neutral mutations and they may eventually also diverge with respect to their phenotype (Figure 10.4).

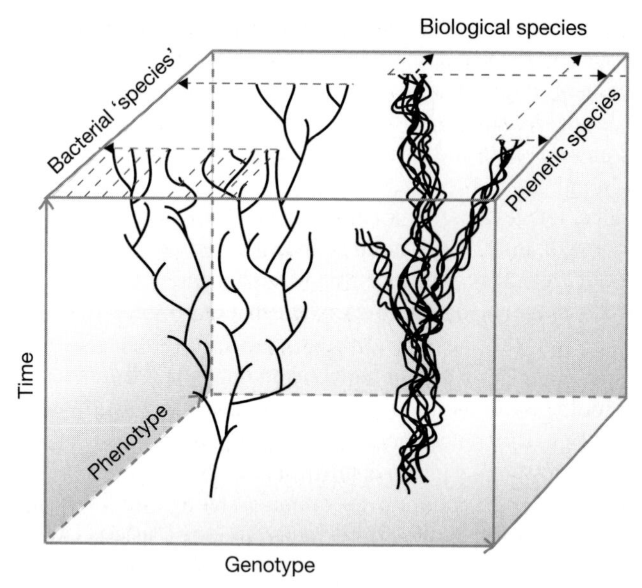

Fig. 10.4 Species and speciation in strictly clonal (left) and sexual (right) organisms. The three axes represent genotype (genetic distance), phenotypic properties, and time. Phenotype is not, of course, a one-dimensional property, but includes a number of dimensions (body size, temperature preference, and so on). Every branch on the trees represents one individual. In clonal organisms individuals are genetically isolated and branching simply reflects cell divisions. Lineages may accumulate mainly neutral mutations, but natural selection may often retain identical or almost identical phenotypes. Different clones belonging to given (phenetic) species may therefore show substantial genetic distances. Occasionally a lineage may evolve into another phenotype. Provided phenotypes are discrete units, species may be characterized, but it is also possible that groups of 'species' sometimes represent a continuum in phenotypic traits. In sexual, out-breeding species (right) biological species represent a common gene pool. There will almost always be genetic differences between individuals, but this variation is constrained by recombination and zygote formation, since large genetic differences inhibit meiosis and zygote formation. Genetic isolation between two populations of a species may lead to speciation from genetic divergence over time and natural selection may lead to phenotypic differentiation.

But this, of course, is not a very useful definition of species. To be sure, horizontal gene transfer does sometimes take place, but chromosomal gene exchanges are probably not common, and like the exchange of plasmids it may occur between unrelated forms. The presence of plasmids in cells is often ephemeral and not of much interest in this context. The crucial question is whether chromosomal recombination is more frequent than mutations and this is probably in general not the case; bacterial evolution may then largely be considered to be clonal.

In traditional bacterial taxonomy species are defined as clonal cultures that share phenotypic traits. Such traits include morphology, ability of producing endospores, motility, types of substrates utilized, types of metabolites, temperature preferences, etc. It was soon recognized that such a definition of species is not absolute and it is to some extent arbitrary; for example it is a common experience in medicine that different strains of a nominal species may be harmless or be pathogenic to a variable degree. Nevertheless, traditional bacterial taxonomy has been very successful, from a practical point of view as in medical diagnostics, but also in describing bacterial diversity. Higher taxonomic levels were also based on phenotypic traits that were believed to be more essential, but it was also recognized that this might not reflect the phylogenetic relationships between different groups of bacteria. Contemporary methods of molecular genetics, that is, nucleotide sequencing of particular genes, totally dominates attempts to unravel bacterial phylogeny, and they have been very successful in this respect (Chapter 12).

The possibility of sequencing genes has also contributed to a rapid and confident identification of bacterial strains. But molecular genetics has not contributed much with respect to defining what a bacterial species is. Sequencing of genes from different strains (clonal cultures) of nominal bacterial species (that are phenotypically identical or almost so) usually results in clusters of genotypes. Nucleotide sequences usually show higher mutual resemblance within strains of a nominal species than they do with sequences deriving from other species. But this has so far only lead to arbitrary species definitions, for example, that in order to be classified as belonging to a species, strains should have 97% identity in the base sequence of the 16S rRNA gene. Similarly the percentage of DNA–DNA hybridization between bacterial strains has been attempted as a criterion for bacterial species. It is striking that the genetic distances between clone cultures belonging to a nominal species are usually very large when compared with interspecies differences among animals. The genetic differences within a bacterial species may be comparable with genetic differences among all mammals.

The reason for these patterns is probably that natural selection tends to maintain certain phenotypic traits, or rather combinations of certain traits, in groups of related (but genetically isolated) bacterial. One can imagine two surviving lineages that resulted from the division of a particular bacterial cell sometime in the past. Over time each of these lineages has accumulated mainly neutral mutations and their genotypes have thus diverged over time. The lineages may have also have diverged phenotypically over time. But it is also possible that natural selection has maintained a particular set of phenotypic traits, keeping the bacteria on an 'adaptive peak', so that they appear phenotypically identical today in spite of substantial genotypic divergence due to differential fixation of neutral mutations in different lineages (Figure 10.4).

Recombination will, of course, counteract this selectively neutral genotypic divergence. But since recombination is so rare (or sometimes, maybe, entirely absent) this may not be very important. A mechanism, recently suggested by F. M. Cohan, may restrict neutral genetic divergence in clonal populations. If a favourable mutation occurs within a bacterial population, then its progeny will quickly out-compete other clonal lineages and consequently purge the population of neutral genetic variation (Figure 10.5). Such a mechanism, if important, could explain the existence of discrete clusters of genotypes that could, perhaps, define species.

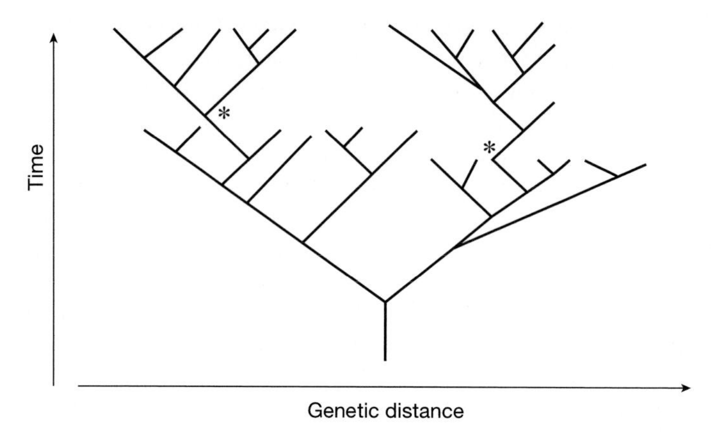

Fig. 10.5 Evolutionary tree for clonal organisms. Favourable mutations are marked by a *. When these occur, other lineages that do not carry the mutation will be displaced through competitive exclusion. This leads to a purging of neutral genetic variation and to discrete clusters of genotypes.

In sexual, out-breeding eukaryotes, divergent adaptive specialization to different ecological niches within a population is constrained by recombination and zygote formation. Adaptive divergence will generally require spatial isolation of subpopulations or some other mechanism that secures genetic isolation. In the predominantly clonal bacterial populations there are no such constraints. Diverging adaptation to particular microhabitats or substrates may therefore readily take place within bacterial populations. Such 'speciation' events may also result from horizontal gene transfer from an unrelated bacterium, thus adding a new phenotypic trait to a cell and its progeny.

Directional selection takes place in bacteria over short periods (even over days if the environmental conditions change). Rosenzweig and colleagues have provided a nice example of bacterial microevolution. Growing *Escherichia coli* in a constant and homogeneous environment with glucose as limiting substrate, it was shown that cells diverged into three geno- and phenotypes. One of these used glucose and the two other types specialized on utilizing different metabolites (acetate and glycerol) produced by the first type. A kind of 'speciation' had taken place, but it would not be useful to name each of these separately. This sort of microevolution probably frequently takes place in nature. But over larger temporal and spatial scales such phenotypic and genotypic changes, or rather fluctuations, generally cancel each other out.

The fact that discrete clusters of genotypes usually correlate with clusters of phenotypes seems to be the rule, and might provide a basis for a bacterial species concept. However, there are still difficulties. Genotype clusters are often nested, meaning that a given cluster, that might define a species, consists of several smaller clusters. Phenotypic differences within genotypic clusters may be slight; in other cases phenotypic differences may be considerable, but genotypic distance may be very small.

Strains of the intestinal bacterium *Escherichia coli* are usually harmless, but some are pathogenic. One of these has even been given a different generic name (*Shigella* causing dysentery) although it falls within the genotypic range of *Escherichia coli.*

The study of microbial microevolution is important in its own right. But the definition of species and nomenclature boils down to a practical problem. Until now a little more than 5000 bacterial 'species' have been phenotypically characterized and named. Recent attempts to extract and sequence DNA from natural habitats such as seawater and soils indicate a genotypic diversity that is much larger than that known from libraries of genotypes of cultured bacteria. Exactly what this natural genotypic diversity means in terms of functional diversity is still somewhat unclear. Certainly, there are undiscovered bacterial phenotypes. But it appears unlikely that there is a huge pool of undiscovered and fundamentally different bacterial phenotypes. In most instances it is the functional diversity of bacteria that is of interest and any practically applicable bacterial species concept must basically rest on their phenotypic properties.

It is likely that some basic evolutionary innovations have occurred in bacteria in the more recent geological past; this of course applies to bacteria that have specialized as symbionts of animals and plants. But generally, most fundamental bacterial phenotypes have probably remained relatively stable over immense periods of time. The quantitative importance of the different microhabitats has changed over geological time, but qualitatively they have to a large extent been similar through most of Earth's history. For example, fossil cyanobacteria from mid-Precambrian time (about 2.5 billion years) are almost indistinguishable from extant relatives (Chapter 13). If time travel was possible then going back in time one could experience exotic menageries of ammonites, dinosaurs, trilobites, etc. To the microbiologist participating in the hypothetical time travel, bacterial biota of the past would look quite familiar and the organisms could readily be named. However, if nucleotide sequencing could be carried out, then substantial differences relative to extant counterparts would be evident.

Large organisms have a rapid species turnover in a geological time-scale. This is caused by frequent local and global extinctions, climatic and geological changes, and isolation of populations and subsequent divergence. Due to their immense absolute population sizes, microbial populations practically never become extinct—neither on a local nor on a global scale. The large population sizes also mean that bacteria can reach any suitable habitat on the globe. Microbial evolution (and ecology) is therefore less influenced by historical contingencies than is the case for macroscopic organisms. As a result many bacterial phenotypes have remained stable over a large part of geological time. Presumably bacteria underwent an adaptive radiation soon after the emergence of the first cells—an event rather similar to the adaptive radiation of animals during the Cambrian. Since the first bacterial phenotypes became established, most have probably remained relatively stable, although some bacteria have adapted to new niches that appeared later on.

The eukaryotic microorganisms take an intermediate position between multicellular organisms and bacteria. Many protist groups are sexless and the species concept (or rather lack of species concept) also applies here. Most advanced groups of protists include sexual representatives; in this case a biological species concept applies—at least in principle.

Geological evidence shows that some large groups of protists arose late during evolution. The diatoms, for example, first make their appearance during the Cretaceous (about 120 million years ago) and altogether there is evidence that the most advanced protists (the 'crown group', see Chapter 12) arose relatively late during the Precambrian. But direct evidence also shows that the evolution of phenotypes has generally been slower than for multicellular organisms, and the explanation for this is similar to that offered for bacteria. The ciliate genus *Tetrahymena* includes about 20 species (of which most, but not all are sexual out-breeders); they are very difficult or impossible to distinguish microscopically. But the sexual forms cannot interbreed and their genetic distances are similar to those of all mammals. The extant species therefore probably go back to the Jurassic. New discoveries of Triassic amber (about 220 million years) contained fossils of a variety of algae, fungi, and protozoa (even including a *Tetrahymena* species). Most of these appear completely identical to extant forms that can be found in any freshwater accumulation today.

When species numbers of different size groups of organisms are compared, then microbes appear surprisingly poor in species when compared with larger organisms. For example, about half a million insects have been named (and some believe that there are many more). In contrast, less than 20 000 species of unicellular eukaryotes are known, and only some 5000 bacteria have been named. Undoubtedly, the species concept used for microbes is not quite comparable with that for animals and plants and many microbes probably still await discovery. Nevertheless, this comparison describes something real. Global species numbers reflect the balance between speciation and extinction. In the case of small invertebrates such as insects, genetic isolation of populations on mountain tops, or on small islands take place continuously and result in speciation. The high global species number therefore reflects a large number of endemic species populations occurring in various isolates. In the case of microbes it *is* the case that 'everything is everywhere'. This means that wherever a suitable habitat is found, then the microbial species that can live there will also be present. Microorganisms are ubiquitous! The capacity for dispersion is usually not due to the per capita capability to disperse, but rather the astronomical population sizes. In one millilitre of seawater, for example, there will be around one million bacteria and perhaps a thousand protozoa. In sediments and in soils these figures must be multiplied by a factor of 100–1000. This means that for purely statistical reasons microbes can reach the most remote places on Earth's surface. It also means, again from considerations of statistics, that local extinctions of species populations practically never occur. Thus species turnover (turnover of phenotypes) is slow over geological time.

Chapter 11

Our anaerobic inheritance

We earlier indicated that life arose under anaerobic conditions. Considerations of geochemistry show this and there is direct geological evidence that the oxygen content of the atmosphere slowly increased over several billion years (Chapter 13).

Even today there are many anaerobic organisms that cannot use O_2 and in many cases they are killed by even trace amounts of oxygen. Anaerobes are found among all three main groups of organisms (eubacteria, archaebacteria, and eukaryotes). In prokaryotes the anaerobic life-style is probably primary in many or most cases although secondary adaptation to anoxia has probably also taken place. In the case of the anaerobic eukaryotic microbes, this is probably in most or all cases a question of secondary adaptation. All green organisms with oxygenic photosynthesis are naturally capable of using oxygen for respiration, although some cyanobacteria can manage anaerobically through fermentation and some can carry out anoxygenic photosynthesis using H_2S as electron donor in sulphidic and anoxic habitats. Animals seem always to require access to O_2, at least during a part of their life cycle. Although our atmosphere now contains about 21% O_2 there are still many completely anoxic habitats. Sediments on the bottom of the sea and lakes are typically anoxic a few millimetres beneath the surface; anaerobic marine basins are not uncommon and the deeper parts of stratified lakes are also typically anoxic. The contents of the intestinal tracts of animals are anoxic and even particles degrading organic matter in the mm–cm range often contain an oxygen-free core.

All extant organisms still carry signs of the anaerobic origin of life. One example of this is that organisms in fact use free oxygen for very few things (if at all). Aerobic organisms use O_2 as a terminal electron acceptor in respiration, but that is about it, at least as far as prokaryotes are concerned. Thus O_2 takes part only to a very limited extent in assimilatory metabolism, and where this is the case it applies to metabolic processes that appear to be of relatively recent origin.

Oxygen is used in the synthesis of sterols that are a fundamental constituent of the cell membrane of eukaryotes. Nevertheless, some protozoa have sterols that do not require O_2 for their synthesis. Free oxygen is also necessary for the synthesis of collagen—a component of connective tissue in animals. It is essential in order that multicellular animals maintain their shape rather than collapsing, and the origin of collagen synthesis is closely related to the origin of animals. The need for collagen is probably one important reason why animals must always have access to oxygen at some time during their life (although a few seem to manage permanently with fermentation for energy generation). 'Quinone tanning' is responsible for the hard parts of the exoskeleton of, for example, arthropods, and it requires access to O_2, but again this is a relative new 'invention' in evolutionary terms. In vascular plants lignin is a kind of analogue to collagen. It is found in complexes with cellulose in the woody parts of plants. These complexes have mechanical strength and are resistant to attack from bacteria and fungi. The synthesis of lignin also requires oxygen, but vascular plants are evolutionarily young. In this connection it is interesting that the microbial degradation of lignin requires an oxidase and oxygen: in terrestrial habitats lignin degradation is carried out first by fungi, but some bacteria also seem to be able to degrade lignin slowly. The necessity for oxygen in lignin degradation is the reason why old wooden ships (such as Viking ships found in Denmark and Norway) can be recovered intact from the sea bottom and why prehistoric wooden tools can be recovered from moorland bogs. If wood consisted only of cellulose, then bacteria would quickly degrade it even under anaerobic conditions.

Oxygenic photosynthesis may have evolved 3.5 billion years ago or earlier. Oxygen as a metabolite had two fundamental consequences for the biosphere. One was the development of the hitherto most efficient type of energy metabolism and this again was the prerequisite for the evolution of all higher life. The other consequence is related to the fact that oxygen is toxic and in a sense the advent of cyanobacteria was responsible for what may have been the largest environmental disaster the biosphere has so far experienced. Even today, all aerobic organisms have an ambivalent relation to oxygen. The fact that health shops are filled with β-carotene, ascorbic acid, and vitamin E as protection against oxygen radicals is, one might say, a trivial example of this; in this case, however, it is largely the pill manufacturers that profit.

Oxygen is toxic for several reasons. A number of enzymes that are particularly important to anaerobes are extremely sensitive to oxygen exposure. Hydrogenase (catalysing $H_2 \leftrightarrow 2H^+ + 2e^-$), nitrogenase (the central enzyme for nitrogen fixation), certain Fe-S electron carriers, and the synthesis of bacteriochlorophylls are all effectively deactivated when exposed to O_2. These enzymes all originated when all life was anaerobic and they have not gained resistance to O_2 since. Oxygen radicals result in a spectrum of more general damage at the cellular level including effects on DNA. Anaerobic microorganisms show considerable variation in terms of tolerance to O_2 exposure. Some have more or less developed versions of the detoxification apparatus found in aerobic organisms. Others, the strict anaerobes, have no defences against oxygen toxicity and they are inactivated or killed by trace amounts of O_2 (concentrations below what can be detected by conventional analytical methods). Such strict anaerobes include, for example, the methanogenic bacteria and the green sulphur bacteria.

Many aerobic microorganisms are 'microaerophiles'; while they utilize O_2 for respiration they seek places with low oxygen tension, typically a few per cent of atmospheric saturation, and in some cases they cannot survive exposure to higher levels. The reason is that protection against oxygen toxicity is energetically costly, and so it is adaptive for these forms to live where only little oxygen is present. Even most of our own cells live in a microaerobic environment, and mitochondria operate at a few per cent of atmospheric O_2 tension.

The O_2 molecule is relatively stable and not in itself very toxic. Problems arise through the incomplete reduction of the molecule. The reduction of O_2 in the electron transport system, catalysed by cytochrome oxidase, is almost complete according to:

$$O_2 + 4e^- \rightarrow 2O^{2-}.$$

But in many other cases of oxygen reduction (spontaneous reactions or catalysed by different cellular oxidases) the reaction follows:

$$O_2 + e^- \rightarrow \cdot O_2^-$$

(where the dot indicates that it is a radical; the molecule has an incomplete outer electron shell and is therefore chemically very reactive). The radical in question is called *superoxide.* It is disputed how dangerous superoxide is in itself. In cells, it is under all circumstances, quickly converted into oxygen and peroxide according to:

$$2 \cdot O_2^- + 2H^+ \rightarrow O_2 + H_2O_2.$$

The process is catalysed by the enzyme superoxide dismutase—metallo-enzymes containing either Cu + Zn, Fe, or Mn. Peroxide is dangerous in the cells because, by oxidizing other compounds: e.g.

$$H_2O_2 + Fe^{2+} \rightarrow \cdot OH + OH^-$$

the hydroxyl radical ·OH is formed. This is considered the most destructive oxygen radical, for example in terms of DNA damage. However in cells, peroxide is quickly degraded to $2H_2O + O_2$ through enzymatic catalysis with *catalase.* Practically all tissue and all aerobic cells contain catalase; this is evident from the fact that, for example, a drop of blood makes hydrogen peroxide foam. Peroxide is also removed through oxidation of various cellular electron donors catalysed by *peroxidases.* In eukaryotes peroxidases occur in special, membrane-covered organelles, the peroxisomes. Superoxide dismutase occurs in large amounts in all aerobic cells and the concentration increases in response to increased oxygen tension. Some anaerobic microorganisms also contain superoxide dismutase (in accordance with the fact that they may occasionally be exposed to some oxygen), but others do not have this enzyme. Catalase seems to occur only in aerobic cells.

There are other mechanisms that protect against oxygen toxicity. Certain compounds such as ascorbic acid and vitamin E react with oxygen radicals. Superoxide also forms photochemically and this is a problem for photosynthetic organisms. Here carotenoids play a protective role.

Oxygen toxicity also has useful aspects. Some blood cells produce superoxide through the process:

$$2O_2 + NADP(H) \rightarrow 2 \cdot O_2^- + NADP^+ + H^+$$

with the purpose of killing invading bacteria. Some plants (St. John's wort, buttercups) have pigments that photochemically produce superoxide for chemical warfare, something that is also known from some ciliates. It is told that in the past, beggars applied extracts of St. John's wort on their skin. After sun exposure the skin blistered and this was supposed to attract the sympathy of by-passers.

Chapter 12

The molecular tree

Principles, assumptions, and problems—the molecular clock

Comparative anatomy is the classical, and still useful, method for unravelling the kinship among organisms. Related organisms, the vertebrates for example, are built according to a common body plan. Certain structures in more or less related organisms are considered *homologous* if they have identical positions in the body plan. Such homologous structures represent different versions of a structure that existed in the common ancestor, but through evolution it has been modified in various ways. Thus the paired anterior limbs of vertebrates (pectoral fins, forelimbs of amphibians, reptiles, and mammals, wings of birds) are homologous structures. The different versions found among terrestrial vertebrates are understood as modifications of the pectoral fins of lobe-finned fish that sometimes during the Devonian acquired an amphibian life-style. All the subsequent modifications of the forelimbs in terrestrial vertebrates (fusion of bones, loss of fingers, etc.) are understood as adaptations to different ways of life, but also as evidence of how evolution has taken place. Such reconstructions of the evolutionary tree are supported by fossil evidence that also provides information on the geological time of the evolutionary events.

When it comes to microorganisms this approach is difficult or entirely impossible. Some groups of unicellular eukaryotes do show many morphological details (for example the ciliates) and then somewhat convincing conclusions on phylogenetic relationships can be made. Electron microscopy has expanded the scope of comparative morphology because finer details can be made out. Chemical criteria, such as types of photosynthetic pigments and the composition of cell walls have also been included. But it has not been possible in this way to account for the relationships between the major groups of unicellular eukaryotes.

In the case of bacteria not many morphological traits are useful and it was recognized early that morphological details were not of much help when trying to resolve the phylogeny of bacteria. In some cases, however, such criteria do delimit particular and (as we now know) natural groups of bacteria. The spirochaetes with their strange motility, and the Gram-positive bacteria with their special cell wall structure are examples. Certain biochemical traits such as chlorophyll *a* in cyanobacteria and methane production in methanogens also seem to define natural groups. Many other such criteria have been shown not to hold. Various phenotypic traits have proved useful for a practical classification of bacteria and for identification. But the phylogenetic significance of these traits has often been unclear.

In the 1960s it became possible to analyse amino acid sequences of proteins. It was shown that proteins with a related function from different organisms also had related, but usually not entirely identical amino acid sequences. With the reasonable assumption that such chemically related molecules reflected their genealogical relationship, it became, at least in principle, possible to unravel the kinship between all kinds of organisms. There are thus homologous genes. The method also had the advantage that differences (degree of relatedness) between organisms could be quantified as genetic distance, that is, the number of nucleotide substitutions when two proteins or genes are compared. In principle it should also be possible to reconstruct the sequence of the protein or gene for a common ancestor.

Comparisons of a given type of proteins from different organisms showed that there are regions within these molecules that appear evolutionarily very conservative whereas other regions are more variable. This is interpreted to mean that there are parts of the protein that are critical to its function and other parts that can be changed through mutations without any effect, or perhaps with adaptive properties.

The earliest investigations of this sort concentrated on animals and plants. This was in part due to the fact that these organisms are closest to us. But it also proved difficult to find convincing homologies in biomolecules from very unrelated organisms (such as bacteria). Vertebrate haemoglobins were popular: they allowed the construction of phylogenetic trees that largely agree with those based on comparative anatomy and palaeontology. It could also be shown that molecular changes appeared to occur at an almost constant rate. If genetic distance between two animal groups was plotted against the age of their common ancestor (known from palaeontology) then a straight line was obtained. Apparently it was possible to use the data as a kind of *molecular clock*, although it had to be calibrated for each type of protein. In principle it should be possible not only to reconstruct phylogenetic trees, but also to determine at what time in the past a branching took place.

The apparently constant rate at which amino acid substitutions take place has lead to the conclusion that most substitutions are selectively neutral. That is to say that the preserved mutations had no or nearly no effect on the fitness of the phenotype, but that they became fixed in populations by random processes (genetic drift) over geological time. In general this is undoubtedly true. There are, however, also examples where substitutions are adaptive. Haemoglobins, for example come in different versions with different oxygen affinities and other properties, and these differences can be related to the biology and habitats of the different species.

More recently it has been shown that there are many cases in which changes in a gene have taken place at different rates in different lineages. Even if the great majority of nucleotide substitutions are selectively neutral one cannot always trust the reliability of the molecular clock. An extrapolation beyond the period from which we have geological evidence is most uncertain. This is regrettable because we are here concerned mainly with evolution that took place before fossils provide unambiguous information on the branching of lineages.

An example of the result of protein sequencing (from around 1970) is shown in Figure 12.1. It shows the evolutionary tree of the respiratory enzyme cytochrome *c*. Cytochrome *c* is composed of about 110 amino acids (there is some variation) plus a

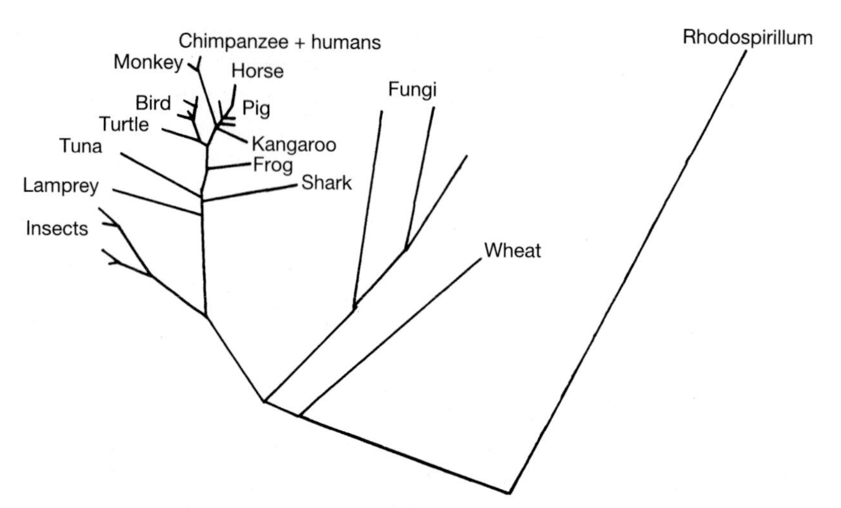

Fig. 12.1 Evolutionary tree for the respiratory enzyme cytochrome c for various eukaryotes and for the purple bacterium *Rhodospirillum*—based on amino acid sequences. Since cytochrome c is a mitochondrial enzyme it is really an evolutionary tree for mitochondria or a tree for the purple bacterium that was a common ancestor for extant purple bacteria and for mitochondria.

porphyrin unit. It is a very conservative molecule that has changed little during evolution. In eukaryotes, cytochrome *c* is a mitochondrial enzyme and Figure 12.1 is an evolutionary tree for mitochondria. *Rhodospirillum* is a photosynthetic purple non-sulphur bacterium. It was included in the study because this group of bacteria have cytochromes that are very similar to those of mitochondria. We know today that this is because mitochondria are descended from representatives of this group of bacteria. The origin of mitochondria is a much younger evolutionary event than the divergence of eukaryotes and prokaryotes. The tree depicted in Figure 12.1 does not therefore reflect the relationship between eukaryotes and prokaryotes, but only the origin of mitochondria and their subsequent evolution.

The lengths of the branches on the tree are proportional to numbers of amino acid substitutions. The cytochrome *c* of humans is identical to that of chimpanzees, but these two species diverged from a common ancestor only some five million years ago. Compared with this the cytochrome *c* of rhesus monkeys deviate by one amino acid; the branching that leads to humans + chimpanzees and to rhesus monkeys took place about 40 million years ago. Marsupials and placental mammals diverged about 120 million years ago and the two groups differ by about 10 amino acids. The tree indicates a 'root' (the break at the bottom of the graph), that is, the position of the common bacterial ancestor of mitochondria and of extant (free-living) purple bacteria. In principle the data on which the tree is based cannot determine such a root (see below); in the figure it is assumed that the evolutionary changes in the cytochrome *c* molecule took place at the same rate in mitochondria and in the bacteria, but there is no evidence that this was necessarily so.

During the 1970s it became possible to determine the nucleotide sequences of genes directly. This was facilitated by the invention of the PCR method (*P*olymerase *C*hain *R*eaction) that allows for the selection and amplification of particular genes (in principle on the basis of a single genome). Since then a variety of genes from thousands of different organisms have been sequenced. In this way it has been possible to construct phylogenetic trees at all scales—from the 'universal tree' and down to the mutual relationships of species within a genus. These methods have proved much simpler than sequencing amino acids in proteins. They also provide more information because several nucleotide triplets code for the same amino acid; nucleotides can therefore, in several cases, be substituted without affecting the amino acid sequence of the protein.

The principle in constructing phylogenetic trees is shown in Figure 12.2. The top of the figure shows (constructed) nucleotide sequences—each with 10 nucleotides—for three species, I, II, and III. It is seen that there is some resemblance, but also differences. There are 30% substitutions between I and II, 50% between II and III, and 70% substitutions between I and III. These percentages measure the *genetic distances* between the species. A sort of phylogenetic tree is indicated; it is constructed so that lengths are proportional to genetic distances.

If there are only three species, only one type of branching (topology) is possible. With four species there are three possibilities (Figure 12.2, middle), and with a larger number of species the number of branching possibilities grows very rapidly. In order

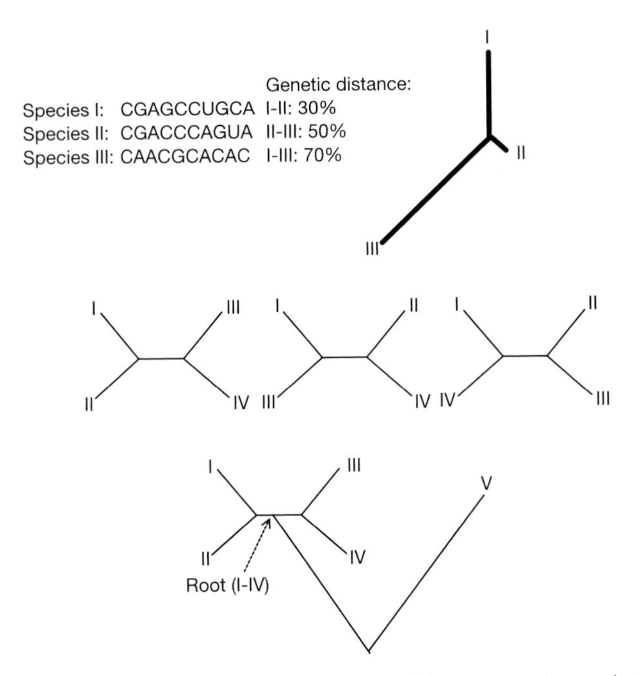

Fig. 12.2 Principle for calculating genetic distance and for constructing evolutionary trees based on nucleotide sequences. For further explanation, see text.

to arrive at the most probable tree (topology and genetic distances) this requires not only data on nucleotide sequences, but also some underlying evolutionary model. Different models are used and they may yield different trees based on identical data sets. Among them is one based on clustering sequences in a hierarchy of highest similarity. Another model, referred to as 'parsimony' calculates the tree that explains the nucleotide sequence data assuming the minimum number of nucleotide substitutions during evolution. In all cases a very large amount of calculation is necessary that must be carried out on a computer. All models can give rise to systematic errors; the method of parsimony, for example, tends to yield internal branches that are too long.

It is noted that the trees in Figure 12.2 (middle) do not have a root—that is the position of the common ancestor for species I–IV. To find this, an out-group is necessary. An out-group means a more distant relative to species I–IV that branched off before the divergence of I–IV. If such a sequence is known (V in Figure 12.2, bottom) then the root for I–IV can be established. The requirement for an out-group would seem to prevent the establishment of a root of the universal tree (that principally includes all living things). However, there is actually a way to solve this as discussed in the following section.

All this sounds—if not simple—then at least something that can be approached systematically, to provide unambiguous results. In reality there are many difficulties, and this applies especially to the earliest branches in the evolution of life. It has already been emphasized that it is only possible to reconstruct the most likely tree. Patterns of substitutions could have arisen in many different ways. For example, the first *C* in the sequence for one of the species in Figure 12.2 could have resulted from the mutation from *C* to *A* and then later back from *A* to *C*, events that contradict the assumption of parsimony. Another problem is that rates of substitutions may differ for different parts of a gene and the substitution rate of given genes may vary in different lineages. Ignorance of such effects—and they are the rule—may lead to incorrect topologies. The form of the trees also depends on the number of sequenced species; use of a higher number of sequences will generally lead to more confident results. Finally, the resolution is limited by numbers of substitutions. The amount of information is limited when there are very few substitutions and also if there are very large differences in sequences. When going far back in evolution it is a problem that a very large number of substitutions have taken place, for example when archaebacteria, eubacteria, and eukaryotes are compared. In such cases information is limited and it may even be difficult to establish homologous parts of a sequence. A number of mathematical statistical methods have been developed that optimize the analysis of sequence data, but many difficulties remain.

Phylogenetic trees for delimited taxonomic categories (e.g. certain animal groups, flowering plants, a particular group of bacteria) do now, to a large degree, reflect reality and such trees are continuously being improved. When it comes to the basal parts of the universal tree the methods have been stretched to the utmost. These cover a long period of evolution with many changes and may reflect events that we can no longer interpret. Under all circumstances, such phylogenetic trees should be considered as hypothesis or models that will continue to undergo changes.

Nevertheless, the universal tree, as it stands today, also has, in fundamental ways, changed our picture of evolution and in certain respects provided robust information.

Before we discuss the universal tree, a final problem should be mentioned; it applies especially to the phylogeny of bacteria. It is known that horizontal gene transfer may happen. If this played a large role during the evolution of bacteria—something we cannot know for certain—then bacteria might be considered only as bags with promiscuous genes. In this extreme case bacteria would not have a phylogeny at all: only individual genes would have an evolutionary history. This interpretation can probably be dismissed. There is after all in most cases a reasonable correspondence between trees based on sequencing of different genes, although it is often not perfect. There are, however, also some data that suggest horizontal gene transfer between archae- and eubacteria during early evolution. A quite similar phenomenon is observed at lower taxonomic levels. When sequences are used to analyse the evolutionary relations between closely related species, or between different populations belonging to a single species, then hybridization or hybridization followed by recombination may require that different parts of the genome must be described by different trees.

The universal tree: archaebacteria, eubacteria, and eukaryotes

Far from all genes show homologies within all living creatures. Some genes may have appeared later during evolution and many have been so changed during evolution that all traces of homologous sequences have been lost. The most conservative genes are those that are related to the most fundamental functions in cells. Among them is the small RNA component of ribosomes (16S rRNA in prokaryotes, 18S rRNA in eukaryotes). Figure 12.3 shows the 16S rRNA of *Escherichia coli.* It consists of about 1500 nucleotides and its shape is determined by internal base pairings. It has proven very useful in phylogenetic analysis because the molecules include some parts that are very conservative and others that have proven more variable through evolutionary time. Sequencing of rRNA has therefore proved useful to estimate relatedness at all taxonomic levels.

In the 1980s Carl Woese was a pioneer in using rRNA sequences to establish the phylogeny of bacteria and to construct a universal evolutionary tree. Since then it has undergone some modifications, but basically it still reflects our current picture of how life evolved—as far back in time as is possible. The tree is shown in Figure 12.4; somewhat more detailed versions for eubacteria, archaebacteria, and eukaryotes are shown in Figures 12.5, 12.6, and 12.7.

The version depicted in Figure 12.4 does have a root indicating the last common ancestor of us all. The nucleotide sequences on which the tree is based cannot alone determine the site of the root. For example, the root might as well have been placed between the archaebacteria and the eukaryotes. The position of the root is therefore based on another set of data. If it is possible to find genes that have undergone doublings prior to the divergence of the three domains (archaebacteria, eubacteria, eukaryotes) then it may be possible to establish the root of the tree.

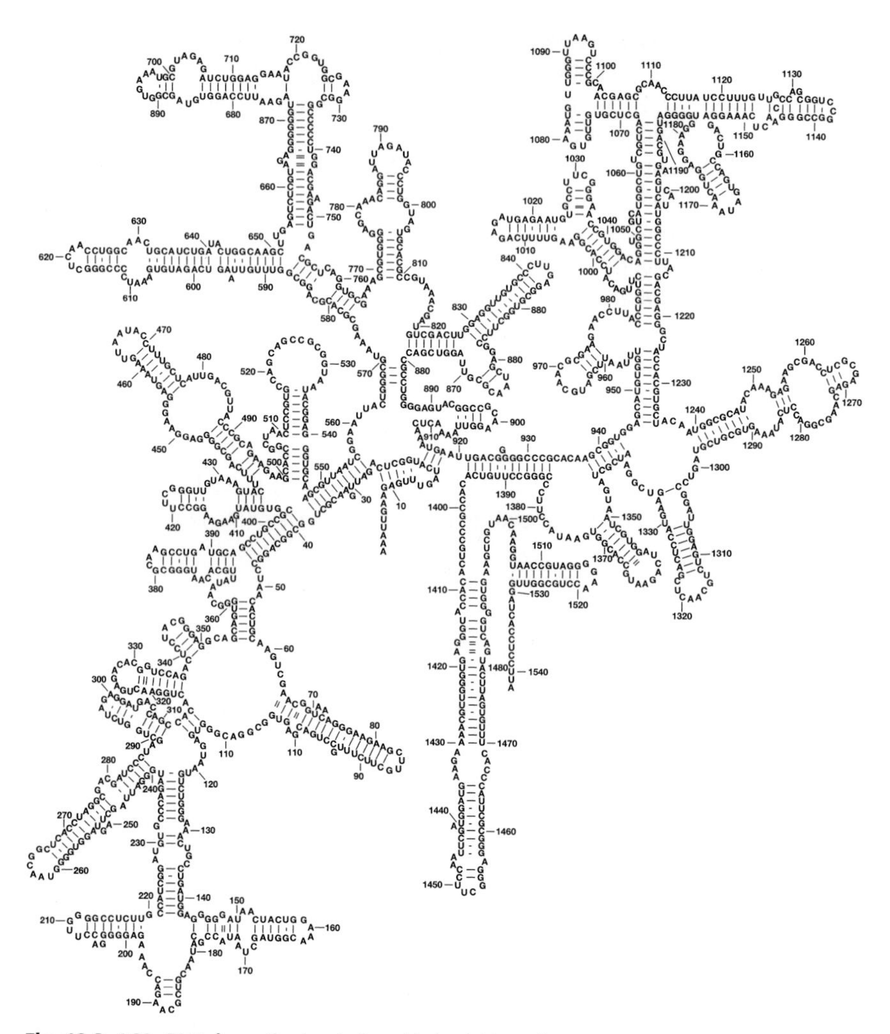

Fig. 12.3 16S rRNA from the bacterium *Escherichia coli*.

With the knowledge of sequences of both genes of all three domains, then it is possible to reconstruct the original (not yet doubled) gene in the hypothetical ancestor. This protogene can then be used as an out-group to place the first branching in the universal tree (Figure 12.5). So far it has been possible to study two such sets of doubled genes. One of these is the pair of genes that specify the enzyme ATPase in all organisms. This pair of genes must have originated through the doubling of a gene in a common ancestor. The other gene that has undergone a doubling in a common ancestor is the so-called elongation factor that plays a role during protein synthesis. In both cases the root appears to be placed as indicated in Figure 12.4 so that the archaebacteria and the eukaryotes are sister groups.

The universal tree allows for some important, in part surprising, and quite robust conclusions. It also implies some difficulties to which we shall return later.

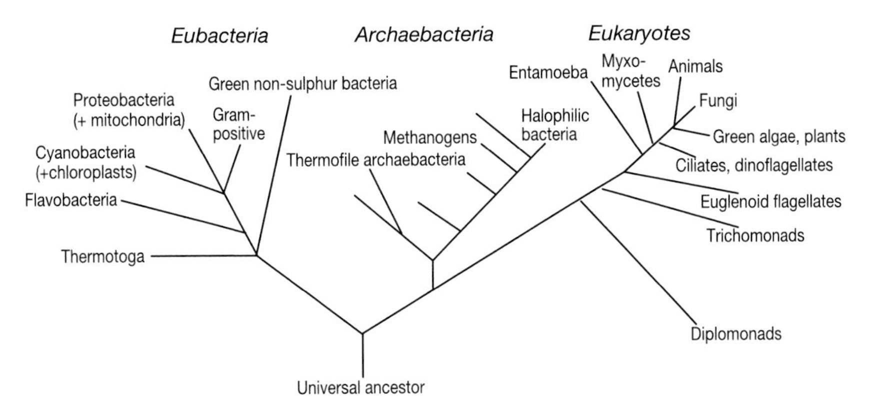

Fig. 12.4 The 'universal tree' based on 16S and 18S rRNA.

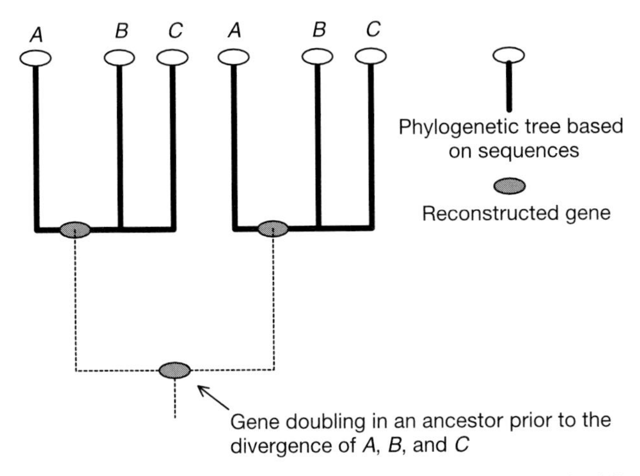

Fig. 12.5 Principle of placing the root on the basis of an earlier gene doubling. On the basis of the two genes it is possible to construct two unrooted trees for the three domains (*A*, *B*, *C*) and the sequence of the original, not yet doubled, gene can be reconstructed. Using this sequence as an out-group it is possible to position the root of the tree.

The first essential result is that all life is related: there is only one kind of life. This has previously been assumed by many people, but not really documented. The second important result is that most evolution seems to be microbial and especially bacterial. Once biology was either zoology or botany. It was assumed that animals and plants constituted the two fundamental groups of living things; various groups of non-photosynthetic protists were classified among the animals and bacteria, fungi, and various photosynthetic protists were considered part of botany. The universal tree shows plants and animals only as tiny twigs at the end of a large eukaryotic branch. One could also look at it in another way, however. It should be recalled that the tree mainly reflects selectively neutral nucleotide substitutions that, over evolutionary time, have been fixed through random processes. The tree may reflect the correct branching pattern. It may also, to some extent at least, reflect the time of various events although there are probably many distortions due to varying rates of nucleotide substitutions in different lineages and due to artefacts introduced by the algorithms employed. But the tree does not reflect phenotypic distances or increase in complexity! The main bacterial lineages diverged very early, but as discussed elsewhere the bacterial phenotypes have probably remained rather stable over billions of years. This probably also applies to the early branches of the eukaryote tree. On the other hand, the tiny twigs that represent animals and plants do not reflect the enormous phenotypic diversification that has taken place within these groups during the last 600 million years or so. In this sense, the classical textbook evolutionary tree (a microbe at the bottom, most detail devoted to animal evolution and a primate at the top) also shows some truth.

The third great discovery is that there are apparently three main groups (domains), and especially that the prokaryotes includes two rather unrelated groups: the archaebacteria and the eubacteria. To stress this point, some have suggested reserving the term bacteria for the eubacteria and to call the archaebacteria archaea. Personally I do not find this very practical. The term 'bacterium' refers to an organizational level of organisms and, for example, the term 'methanogenic bacteria' is well established (and it will cause confusion in academia if scientists working on these creatures call themselves 'archaeologists'). In many important respects the archaebacteria are typical bacteria in terms of structure and function even though the term bacteria does no longer describe a monophyletic group. The term archaebacteria is due to Woese and implies that these organisms are especially primitive or primordial. The form of the universal tree does not really support this notion. There is also some feeling that the apparent large genetic distances between the domains are an artefact. Certainly, they cannot represent a constant molecular clock unless the common ancestor lived before the origin of Earth. If, for example, the animal branch corresponding to 600 million years is used as a scale, then the divergence of the domains took place 4.8 billion years ago!

The differences between the domains have already been considered to some extent. The eubacteria (Figure 12.6) represent the most varied group as far as energy metabolism is concerned. Photosynthesis occurs in five of the main branches with 'deep roots' (green non-sulphur bacteria, green sulphur bacteria, Gram-positive bacteria, cyanobacteria, and proteobacteria). Oxygenic photosynthesis occurs only among the cyanobacteria that show some relation to the Gram-positive bacteria; in all other groups only anoxygenic photosynthesis is found. The proteobacteria represent a core group among the eubacteria. They can be divided into five main groups (α-, β-, γ-, δ-, and ε-proteobacteria). Three groups include photosynthetic forms (purple non-sulphur and purple sulphur bacteria), but these also include O_2 respiring forms that oxidize organic substrates, reduced sulphur or iron compounds, or ammonia. One group includes most of the sulphate respiring bacteria. Many trivial and well-known bacteria such as the enterobacteria including *Salmonella*, *Escherichia coli*, and some other, in part pathogenic bacteria also belong to the proteobacteria. The α-group includes the nitrogen-fixing symbionts (*Rhizobium*) of legumes, species

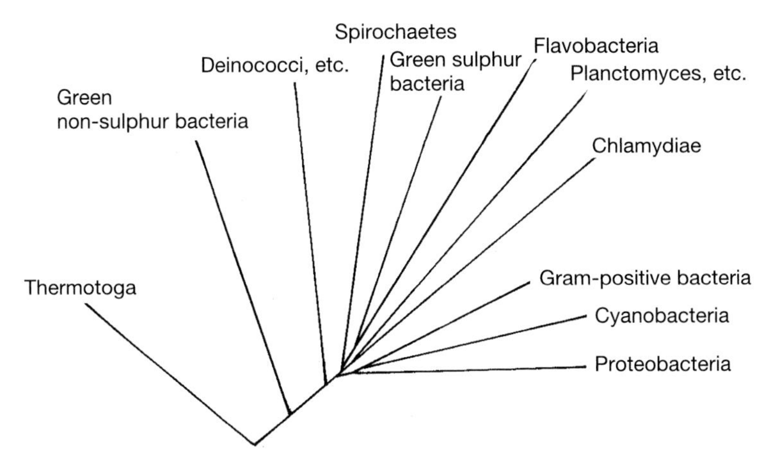

Fig. 12.6 16S rRNA tree for the eubacteria.

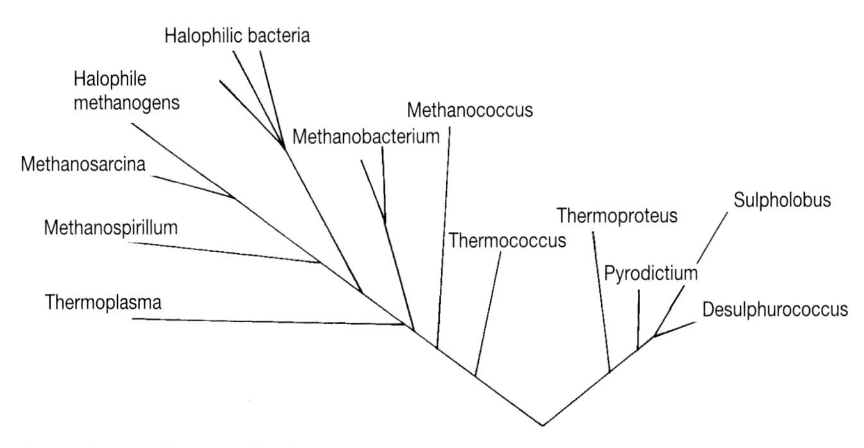

Fig. 12.7 16S rRNA tree for the archaebacteria.

that can respire using nitrate as electron acceptor, several photosynthetic forms, and the ancestors of our mitochondria. The proteobacteria represent a good example of how the different types of energy metabolism could change rapidly over evolutionary time through minor modifications of a few enzymatic components.

The eubacterial tree includes a dilemma. Geological evidence (Chapter 13) seems to indicate that cyanobacteria arose at least 3.5 billion years ago. On the other hand, the rRNA tree does not indicate very deep roots for this group and it would seem that the cyanobacteria represent a relatively young lineage among the eubacteria. It is possible that the molecular clock has moved more slowly within this group; at least it does appear probable that most other bacterial lineages had already diverged prior to the advent of the cyanobacteria.

The archaebacteria (Figure 12.7) deviate from the eubacteria with respect to nucleotide sequences of rRNA and other basic genes. But they also have a deviant cell membrane and cell wall and they have introns in their genome. Genes with eubacterial-type nucleotide sequences occur in some archaebacteria—this has lead to the assumption that horizontal gene transfer has taken place across domains early during evolution. There are three main groups of archaebacteria. The CH_4-producing methanogenic bacteria have a unique type of energy metabolism (Chapter 7); they occur in almost all types of anaerobic habitats. The extreme halophilic bacteria can live in concentrated brine and they have an absolute requirement for high ambient Na^+ concentrations. They are basically oxygen respiring organisms that use organic substrates, but they also have a unique type of phototrophy. The extreme thermophilic bacteria are sometime considered as a special subdomain (Crenarchaeota). They are especially known from hot springs, hydrothermal vents, and other hot habitats. Their energy metabolism is mainly based on sulphur compounds or respiration based on metallic electron acceptors, but some also respire oxygen. As already discussed (Chapter 8) they share a few traits with the eukaryotes.

The fundamental properties of the eukaryotes were discussed in Chapter 8. Their very deep root (Figure 12.4) is striking; it may reflect the apparent discontinuous gap to the prokaryotes, or it may be an artefact. Under all circumstances it is likely that their evolution goes further back than the 1.5–2 billion years from when the first fossil remains suggest their presence and from when molecular data indicate the origin of mitochondria.

The eukaryote tree in Figure 12.8 shows only some of the many protist lineages. It is peculiar in that there are relatively few lineages with very long roots and then an apparently more recent divergence into many lineages: ciliates, dinoflagellates, apicomplexa (parasitic protozoa including, for example, the malaria parasites), heterokont algae (several unicellular photo- and heterotrophs, brown algae), red algae, green algae, plants, fungi, and animals. These lineages are often collectively called the 'crown group'. They represent the vast majority of the eukaryotes in the extant biosphere. It is conceivable that they evolved late towards the end of the Precambrian during a period when the atmosphere had reached a certain O_2 tension. Plants, animals, and fungi are relatively closely related, but they derived independently from different protists. The fungi are more closely related to animals than to plants (which should be noted especially by botanists and vegetarians).

The diplomonads (and presumably another group of amoeboid forms) seem to have diverged earlier than any other extant eukaryotes and they are the most primitive known representatives of the domain. They are obligate anaerobes and lack mitochondria and an endoplasmic reticulum, and sexuality is unknown. Some enzymes are more closely related to counterparts among the eubacteria than to those of other eukaryotes. The absence of mitochondria has until recently been considered original, but the presence of a mitochondrial-type chromosomal gene has challenged this interpretation. All the other early eukaryotic lineages are represented by amoeboid or flagellated protists; they all have mitochondria or organelles that can be derived from mitochondria, and sexual processes are known in some cases.

The difficulties when trying to derive eukaryotes from prokaryotes were discussed in Chapter 8. It should be mentioned that when nucleotide sequences of enzymes

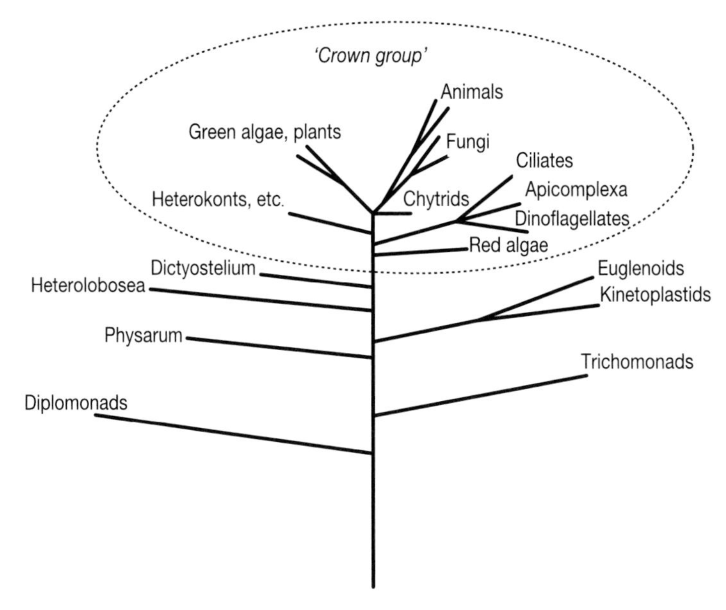

Fig. 12.8 18S rRNA tree for the eukaryotes. The shorter branches on the top (including the red algae) are referred to as the 'crown group'. They seem to reflect an adaptive radiation that took place relatively late during the Precambrian.

are compared, those involved in energy metabolism do in many cases appear to be most related to counterparts among the eubacteria (not considering chloroplast or mitochondrial genes, that are obviously of eubacterial type). In contrast, genes that are involved in the genetic apparatus and in transcription and translation are more similar to corresponding genes of archaebacteria. This has lead to the hypothesis that the protoeukaryote arose as a chimera between an archaebacterium and a eubacterium so that the cell nucleus corresponds to a symbiotic archaebacterium living inside a eubacterium. It is probably the model that for the time being best explains the molecular facts, but it is also rather speculative.

The universal ancestor

It is fashionable to 'seek ones roots'—so who was our common ancestor? The universal tree (Figure 12.4) suggests substantial genetic distances between the domains. This fact and the desire to understand how the first cell arose has lead to some wild speculation about just how primitive the universal ancestor was: that it represented a stage prior to the evolution of a real genome (a 'progenote') and even that it still lived in an RNA world.

A more sober approach is first of all to recognize that the most striking feature of life is its similarity! All really fundamental cell functions (the genetic code, transcription and translation apparatus, the essential components of energy metabolism, the functional and structural properties of the cell membrane) are common to all organisms that exist today. This provides clear boundary conditions for any further speculations. On the basis of the common properties of the three domains (or at least of the eubacteria and the archaebacteria) it can be established that the universal ancestor had the following properties. First of all it had a cell membrane with similar functions to extant cells (energy transduction, transport proteins, etc.). We may guess that its basic building blocks were phospholipids as in eubacteria and in eukaryotes and that the isoprenoid cell membrane of archaebacteria is a derived feature, but we cannot know for sure. The ancestor may have had a cell wall, but the difference between the versions found among eubacteria and archaebacteria makes this uncertain. The ancestor had a genetic system based on DNA, a single circular chromosome, and transcription and translation took place as in extant prokaryotes. Energy transduction was based on a proton (or Na^+) pump in the cell membrane and an ATPase coupled to the return flux through the cell membrane. Altogether: a description of a generalized bacterial cell as we know them today!

When it comes to a more detailed specification of the type of energy metabolism we are on considerably more uncertain ground. Much speaks in favour of a type of light-based energy metabolism because phototrophy is represented among several eubacterial lineages with deep roots and because respiratory energy metabolism can easily be derived from phototrophy. Intuitively it seems likely that electromagnetic radiation from the sun was from the beginning, the ultimate energy source when molecules became life.

Through the analysis of nucleotide sequences of genes we have reached a horizon behind which we cannot see. The universal molecular tree has brought us back intime, perhaps about four billion years when Earth had only one-eighth of its present age. At this time bacteria had already originated and they already shared all their basic properties with extant bacteria. The fact that we can draw a profile of the common ancestor for all extant living things is, after all, a kind of triumph. It is possible that future clever analysis of molecular data can push this horizon somewhat back in time and it is also possible that one day extant representatives of a hypothetical fourth domain will be discovered that can throw some more light on life's earliest evolution.

Chapter 13

Evidence from geology

Earth's active surface

During the 1960s geology obtained a coherent description of structures and processes in the surface layers of Earth. These theories are collectively termed *plate tectonics*. Radioactive decay and thus heat generation in the mantle leads to slow convective flow somewhat comparable with that in a viscous fluid exposed to heat from below: hot fluid rises in some places from the bottom to the surface, cools at the surface to sink down in other places. In a similar way molten magma rises in certain zones below the oceanic sea bottom and spreads to the sides as new ocean bottom. These spreading zones show through mid-oceanic ridges, volcanism, and hydrothermal vents. Hydrothermal vents occur where seawater penetrates down in the ocean bottom trough cracks close to mid-oceanic ridges. The water is heated and reappears at the surface containing sulphide, methane, hydrogen, and metal ions deriving from chemical reactions between the rocks and water. Iceland, for example, is situated on the mid-Atlantic ridge and over geological time the island will split into an eastern and a western part.

The floor of the oceans consists of tectonic plates; altogether there are seven larger and a number of smaller ones on the surface of Earth. These tectonic plates move away from spreading zones at a velocity of a few centimetres per year. The continents consist of the lightest mineral components on Earth's surface. They float on the top of the uppermost layers of the mantle and they are in continuous movement due to ocean bottom spreading. Where the tectonic plates collide with continents the ocean floor is pressed down into the mantle again. These *subduction zones* form deep sea trenches in front of the continents. They also cause mountain building, volcanism, and earthquakes (Figure 13.1). The Pacific tectonic plate is squeezed beneath the west coasts of North and South America resulting in the Andes and Cascade mountains as well as volcanoes and earthquakes. Subduction also means that sedimentary deposits on the ocean floor are transported down into the mantle. Geologists believe that the continents have grown during geological time. The positions of the continents on the surface of the Earth are fairly well known for the Phanerozoic period (the last 544 million years). For example, they were all assembled in a single super-continent about 250 million years ago. They then spread apart again, first in a northern and a southern continent. Then the Atlantic Sea opened during the Cretaceous period and the southern continent broke up into what is today Africa, India, Antarctica, and Australia. During the Tertiary, India collided with Asia resulting in the Himalayas and Africa moved north resulting in the formation of the Alps and the Pyrenees. The position of the continents has an effect on the global climate

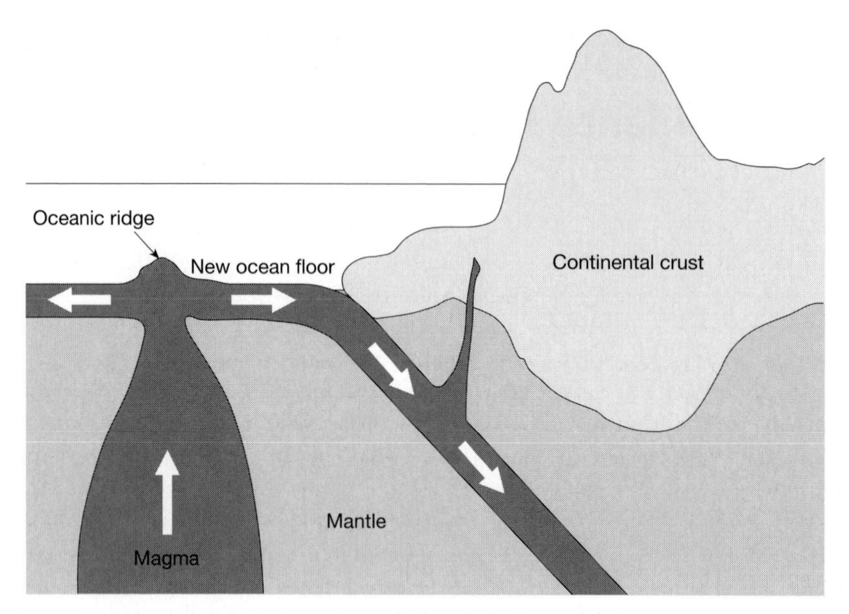

Fig. 13.1 Processes resulting from plate tectonics. For further explanation, see text.

by determining the course of ocean currents. Increased volcanism following more dramatic tectonic events also affects climate by releasing large amount of CO_2 to the atmosphere.

One consequence of plate tectonics is that the ocean floor is young: it is constantly formed at spreading zones to become destroyed again by subduction. The oldest known ocean bottom is little more than 200 million years old; sedimentary deposits on the ocean floor can therefore not provide any evidence on the history of life earlier than the Jurassic. In order to study older events only sediments formed in continental seas (on the continental shelf and typically at depths less than about 200 metres, including very shallow deposits and deposits in river deltas and lakes) are available. Sedimentary deposits on continental shelves may also be squeezed down into the mantle in connection with subduction zones. They may also be pressed up during mountain building to become eroded and lost. Nevertheless, some sedimentary rocks have been preserved even from the early Precambrian. Access to such deposits depends on several facts. The current sea level is relatively low (since much water is now bound in the Arctic and Antarctic ice-sheaths). Many sedimentary deposits have been pushed up into mountain ranges, and older deposits have subsequently become exposed and available for study through erosion of younger, overlying deposits.

Newly formed mountains immediately start to erode due to the chemical and physical actions of the atmosphere. Rocks erode mechanically through frost and temperature fluctuations and the resulting mineral grains (sand and clay minerals) are washed out via rivers to deposit on the sea bottom. Chemical erosion results from the atmospheric content of CO_2 and water so that calcium and magnesium silicates (that constitute a large part of continental crust) dissolve into Ca^{2+}, Mg^{2+}, HCO_3^-, and SiO_2. The ions end up in the sea and deposit as carbonate rocks (calcite, dolomite). The presence of oxygen and water leads to the oxidation of newly exposed minerals that contain reduced sulphur and ferrous iron.

Biological processes contribute to erosion in various ways. They also contribute to the formation of sedimentary deposits on the sea bottom. Thus carbonate rocks are to a large extent formed through the accumulation of skeletal components of plankton organisms and with contributions from mollusc shells, corals, etc. Other organisms (diatoms, radiolaria) are responsible for silica-containing deposits.

Sedimentary rocks that have not been exposed to very high temperatures also contain organic matter (*kerogen*). Most particulate organic matter that reaches the sea bottom is rapidly mineralized due to the activity of bacteria and other organisms, but a small part is not broken down and it accumulates in the sediment. This material consists mainly of compounds such as hydrocarbons, lipids, humic substances, and lignin that are broken down slowly or not at all under anaerobic conditions. Under suitable high temperatures and pressure these compounds may transform into hydrocarbons (crude oil, methane) that sometimes accumulate in particular geological formations and can be exploited as fossil fuels. Freshwater swamps are anaerobic as well as acid, and organic matter tends to accumulate in the form of peat. Over geological time, peat is transformed non-biologically into lignite, and eventually into coal and anthracite. Organic remains in sediments can, if they have not been greatly transformed by heat, provide information on the organisms that produced them.

Tectonic processes, mountain building and subsequent erosion, and volcanism result in recycling of materials that would otherwise have been lost to the biosphere. This has been a decisive mechanism for the origin and maintenance of life on Earth, for these processes have ensured that the original atmospheric contents of CO_2 has not entirely disappeared in the form of carbonate rocks, or been buried as organic matter. Likewise, erosion of exposed crust has secured a recycling of soluble phosphate that would otherwise have accumulated as inaccessible Ca and Fe phosphates.

Speculations on properties of the primordial atmosphere

The oldest known rocks on Earth have been dated to almost four billion years. These are crystalline continental rocks, showing that some continental crust had already formed then. The slightly younger (about 3.8 billion years) rocks from Isua in Western Greenland include sedimentary rocks although they are rather metamorphosed. This shows that oceans and a hydrological cycle already existed at that time. Assuming the pressure of the atmosphere was like it is today, this also means that temperatures were in the range 0–100 °C—something that is supported by indirect evidence of life in these deposits. The period between the formation of Earth (about 4.6 billion years) and the formation, 600 to 800 million years later, of the above mentioned rocks has left no trace. All considerations of the earliest physical and chemical conditions of the surface of the Earth are therefore speculative and unfortunately there are very few constraints on such speculations.

It is commonly believed that the Earth's atmosphere resulted from volcanic out-gassing, although it has also been suggested that impacts of comets could have contributed volatiles such as water. The atmosphere is thus secondary in the sense that during the formation of the Earth through accumulation of mineral grains and other solid objects it could not hold on to gasses in the solar nebula. Atmospheres of small planets therefore contain much less hydrogen and helium than is the case for the sun and the large planets.

Today, volcanic gasses consist mainly of N_2, H_2O, SO_2, CO_2, and some H_2; that is, an O_2-free, but not very reducing mixture. If this was so for the early atmosphere of the Earth, it causes difficulties for understanding the synthesis of organic molecules as in Miller-type experiments (Chapter 3) although it is not entirely impossible. The primordial atmosphere could also have been more reducing. This could have happened if the formation of the Earth took place relatively quickly (over 10–100 million years). This would mean that the surface of Earth would have been hot and molten, and metallic iron that had still not sunk to the core, could act as a reductant and produce an atmosphere rich in H_2, CO, and H_2S. Photochemical formation of CH_4 and NH_3 would then be possible from H_2, N_2, and CO. Under all circumstances it must be assumed that volcanic and other geothermal processes were more frequent and intense during the early history of Earth. In addition to the heat released through meteorite impacts, the more intensive radioactivity also produced more heat. Extrapolating back from the extant amounts of natural radioactive isotopes and their known half-lives, it can be calculated that heat production in the early Earth was some six times greater than it is today.

It is agreed that O_2 could occur only in trace amounts; exactly how much is more debatable, but it could probably not have been maintained above about one-millionth of the extant partial atmospheric pressure. Some oxygen would have been formed photochemically in the upper atmosphere, but it would also quickly combine with reduced iron and sulphur resulting in an extremely low oxygen tension. The light hydrogen molecules, formed through photochemical processes, would have a tendency to evade the gravitational pull of Earth and escape to space. Thus even if life had never evolved, the atmosphere and the surface of Earth would slowly have become more oxidized and have developed into something similar to what is found on Mars: an almost oxygen-free atmosphere, but with an oxidized surface. The continuing geological activity on the Earth would, however, probably have counteracted this to some extent.

It is assumed that the atmospheric CO_2 concentration was much higher during the Earth's early history and that slowly, if irregularly, it decreased to the present level of about 0.03%. Astronomers believe that the energy flux from the young sun was about 30% lower than it is now, but that the higher atmospheric CO_2 content, through a greenhouse effect, maintained a warm or temperate climate. A hypothetical higher atmospheric methane concentration may also have contributed to a greenhouse effect. The earliest signs of glaciations are a little more than two billion years old. During the last 900 million years, ice-ages appear to have been more frequent than earlier in Earth's history (Figure 13.2). Some people assume that early Precambrian climate was hot, but in fact this is not known.

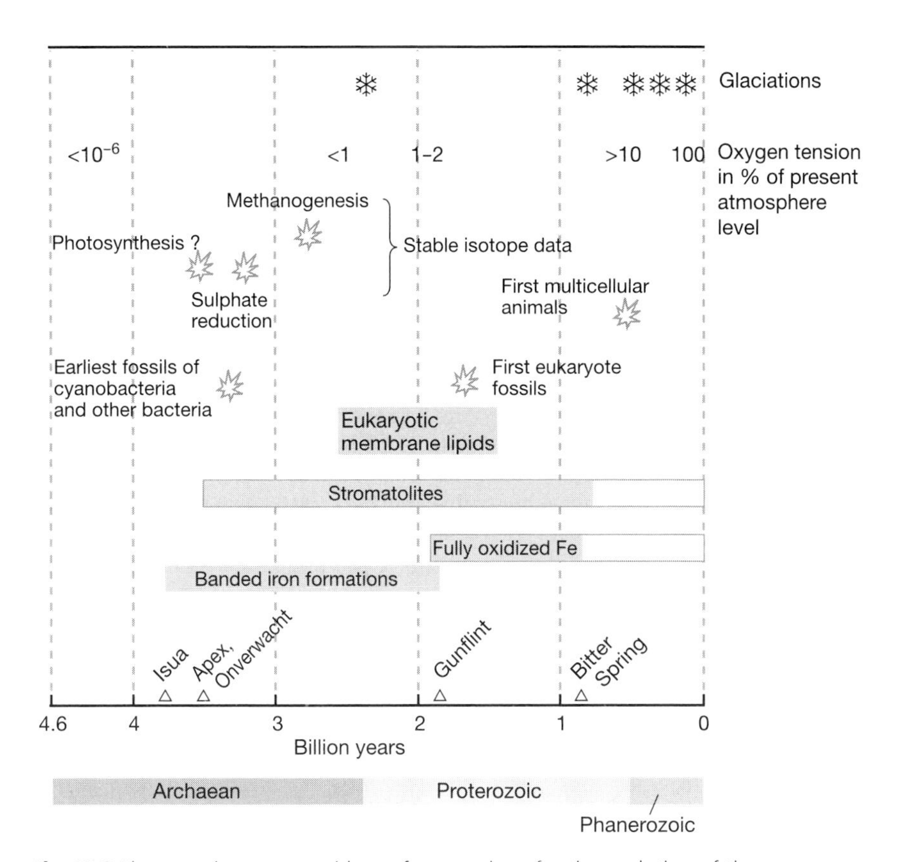

Fig. 13.2 The most important evidence from geology for the evolution of the biosphere during the Precambrian.

The nature of geological evidence for early life

We have more or less direct evidence of life through the last 3.8 billion years of the Earth's history. This evidence stems mainly from sedimentary rocks deposited in coastal (continental) seas and to a somewhat lesser degree in lakes and on land. As already discussed it is thanks to tectonic processes in conjunction with erosion that we can find such remains. But these processes have simultaneously destroyed such remains over geological time. Many sedimentary rocks have been lost entirely through erosion or they have been metamorphosed by heat and pressure leading to melting and re-crystallization. It may finally be speculated that some well-preserved remains still exist, but that they are inaccessible (e.g. beneath the ice-sheet of Greenland). The number of accessible and relatively unaltered sedimentary rocks decreases approximately exponentially with geological age and the very oldest ones (>three billion years) are very few. Yet, some of these have given valuable information that to some extent constrains speculation.

The finding of real fossils of microbes is perhaps the most convincing and informative evidence that we have. Due to their cell walls, bacteria may become fossilized under special conditions so that they retain their original (if somewhat compressed) shape. This can happen when the bacteria become embedded in amorphous silica that eventually turns into cherts (flint-like siliceous rocks). Unfortunately, the morphology of bacteria rarely tells anything about their functional properties. The cyanobacteria constitute an exception due to their extraordinarily large sizes and the characteristic colonies formed by some species. Cyanobacteria are among the oldest recognizable fossils known. *Stromatolites* (that in most cases do not harbour recognizable fossils) are remains of microbial mats dominated by cyanobacteria. They represent a type of microbial community that must have dominated shallow water habitats throughout most of the Precambrian. Eukaryotic cells that lack cell walls are less prone to leave traces in the form of recognizable fossils. However, many do produce some sort of cell wall, test, or other skeletal components consisting of silica, carbonate minerals, cellulose, or chitin-like material. Traces of such organisms are known from the latest one billion years of the Precambrian.

Organisms also leave traces in the form of organic compounds that in some cases are diagnostic of particular groups of organisms. There may, however, be considerable difficulties in proving that organic matter extracted from Precambrian sedimentary rocks actually derives from contemporary organisms. Organic compounds of biological origin are omnipresent. The problem is not only one of laboratory contamination, but dissolved organics may have entered rocks via cracks during some later geological period. Several earlier reports claiming the presence of particular organic compounds in geological formations have later been questioned. The presence of kerogen in even the oldest sedimentary rocks is considered bona fide evidence of contemporary life. The characteristic membrane lipids of eubacteria, archaebacteria, and eukaryotes have left identifiable hydrocarbon derivates in sedimentary rocks; these and fossil porphyrin structures are now also believed to represent reliable evidence of contemporary origin.

The mineral composition of certain types of deposits provides some information on the contemporary chemistry of the atmosphere and seawater. Far from all aspects can be deduced this way, but some key stages in the evolution of atmospheric oxygen contents have been determined.

The last type of evidence depends on stable isotope composition of carbon and sulphur compounds. Both elements have each two stable isotopes: C-12 and C-13 and S-32 and S-34, respectively. In both cases the lightest isotopes are by far the most common ones. The light and the heavy isotopes are chemically similar, but it requires slightly less energy to break bonds in compounds containing the lighter isotope. Photosynthetic organisms therefore discriminate between $^{12}CO_2$ and $^{13}CO_2$ in favour of the lighter isotope during photosynthesis and the organic matter that is produced is therefore enriched with the lighter isotope of carbon relative to the external inorganic carbon. Such discrimination will (in the case of the carbon-fixing mechanism found in cyanobacteria and purple bacteria) lead to an enrichment of about 1.8% for C-12, whereas the remaining carbonate is deprived by about 0.07% of this isotope. Such a differential isotope ratio between organic matter and contemporaneously deposited carbonate minerals indicates that the organic carbon was derived from photosynthesis. In practice an isotope standard is used and it is sufficient to know only the stable isotope ratio of organic matter. The oldest (3.8 billion years) sedimentary rocks known from the Isua formation contains graphite granules. They show enrichment in C-12 that is consistent with the assumption that the graphite represents remains of photosynthetic organisms (though not necessarily oxygenic photosynthesis). There is thus evidence that bacterial plankton with some sort of autotrophic C-fixation was already established by then. Such C-13 depletion of organic matter is also known from 3.5 billion years ago and up until present, in all cases probably mainly reflecting oxygenic photosynthesis.

Methanogenic bacteria show an even higher degree of discrimination against C-13 so methane that is formed is C-13 depleted by 6–8%. The methane is not directly deposited, but it can in part be utilized by other organisms as a carbon source and thus the produced organic matter is C-13 depleted. The oldest sign of methanogenesis based on stable isotope rations goes back only to about 2.7 billion years. It is generally believed that methanogenesis is a much older biological process, but that methodological difficulties have not allowed its detection in older deposits.

Sulphur isotopes can similarly be used to demonstrate sulphate respiration. In this process the principal metabolite (sulphide) is enriched with S-32 by about 4% relative to the isotope ratio of sulphate. Such a difference in the isotope ratio of contemporaneously deposited anhydrite (gypsum, $CaSO_4$) and pyrite (containing reduced sulphur possibly of biological origin) would then indicate sulphate reduction. Most recently, sulphate reduction has in this way been demonstrated in 3.5 billion-year-old deposits.

Interpretation of stable isotope data is not always straightforward, and in fact it suffers from several difficulties. The expected isotope distributions assume an open system. If, for example, it is assumed that the seas once contained only small amounts of sulphate, or that sulphate reduction took place in closed systems with a small sulphate pool and the sulphate was consumed as fast as it was produced, then isotope discrimination would be less pronounced or absent. This is because most of the sulphur present would at any time be in the form of sulphide irrespective of stable isotope composition. This may be one reason why it has been so difficult to detect sulphate reduction in the earliest Precambrian deposits. The stable isotope method is an important tool—but like almost all other evidence of Precambrian life, interpretation is tricky. The major events and methods for investigation of Precambrian life are summarized in Figure 13.2.

Precambrian fossils and Precambrian microbial communities

There is evidence of Precambrian planktonic communities, but from the earliest periods this is largely in the form of organic matter in sedimentary rocks. Until about two billion years ago plankton communities included only bacteria and it may be assumed that unicellular cyanobacteria were important in the photic zone of the seas—as they are today. Not much more can be said about early Precambrian plankton communities.

When it comes to benthic shallow water communities we are a whole lot better off. The most widely distributed and obvious remains are the so-called *stromatolites* (Figure 13.3; see also Plate 4). Stromatolites are laminated structures consisting of carbonate rocks (limestone, dolomite) but silicified stromatolites also occur. They have been known for a long time, but the interpretation of these structures is more recent. Today most stromatolites are believed to be the remains of microbial mats, that is, communities of bacteria and especially filamentous cyanobacteria. These mats must have been the dominating biological communities on the bottom of shallow seas and along shorelines over a period of about three billion years (about 65% of Earth's history!). The earliest known stromatolites are about 3.5 billion years old. From about 600 million years ago and up until the present, they occur

Fig. 13.3 (See also Plate 4) (Top) A 2.2. billion-year-old stromatolitic rock from Belfast, Transwaal (South Africa). (Bottom) Thin section of an approximately two billion-year-old silicified Gunflint Formation, Canada, showing fossils of cyanobacterial filaments and of other microbes. (S. M. Awramik)

more rarely in the geological record. But they actually still exist in special habitats where animals are absent or rare; otherwise grazing and mechanical disturbance prevents their development. It can therefore be assumed that the advent of multicellular animals almost exterminated these biological communities that had otherwise dominated over such a long part of Earth's history. Extant cyanobacterial mats will be discussed in the following section. In many essential ways they are true analogues of Precambrian stromatolitic mats. As such they may yield information on structure and function of their Precambrian counterparts, and provide a picture of how the shallow sea bottom and the shoreline looked from 3.5 billion years ago until animals arose some 600 million years ago (Figure 13.4; see also Plate 5).

Fossil stromatolites are found on all continents—and in some places they are even quarried as building material. In most cases they do not contain bacterial fossils, but various characteristic patterns in the laminated structures can be interpreted in the light of extant microbial mats. The laminations are usually interpreted as seasonal patterns in growth or seasonal patterns of sedimentation of mineral particles on the top of the mats. Recent experimental observations indicate that laminations in carbonate deposition may develop even under constant external conditions.

In rare cases bacterial fossils occur in stromatolites and also in contemporary cherts. The shape of the cells and to some extent their internal structures have been preserved to a degree that is astonishing. This presumably happened when the cells became impregnated by amorphous silica that later hardened to become a chert. Today, this phenomenon is rare (it is known from some mats growing in hot springs such as in Yellowstone Park). When the cherts are ground into thin slices and polished it is possible to study the fossils microscopically and often also to see how they were originally oriented within the microbial mat. Such fossils demonstrate beyond doubt the nature of stromatolites and the presence of cyanobacteria far back in the Precambrian.

The oldest recognizable microfossils are 3.5 billion years old and thus contemporary with the oldest known stromatolites. These derive from the Apex chert from Western Australia and from Onverwacht from Southern Africa. A classical finding is the somewhat younger Gunflint Formation in Canada (Figure 13.3; see also Plate 4) and the even younger Bitter Spring in Australia has provided extremely well-preserved fossils. Fossil-containing Precambrian stromatolites and shale are also known from other places such as China, Russia, and USA.

The fossils include bacterial cells for which it is impossible to say anything about their properties or taxonomic position. They also include fossils of filamentous cyanobacteria. In particular, those that are younger than 2.7 billion years have a striking similarity to certain extant cyanobacterial species, from which they would probably be indistinguishable of the fossils if they could be brought back to life. J. W. Schopf has measured various dimensions such as cell diameter, the ratio between cell diameter and length, etc. In all cases these coincide with similar measures of extant forms. There cannot be any doubt that these fossils represent cyanobacteria and that even the phenotypes are probably similar to extant forms. The oldest (>2.7 billion years) fossils seem to deviate from extant forms and some scientists think that they were not cyanobacteria, but some other form of filamentous bacteria.

In particular, the nature of the 3.5 billion-year-old Australian fossils has recently been questioned and it has even been suggested that they represent non-biological structures rather than bacterial remains. Circumstantial evidence of contemporary O_2 production does, however, suggest that cyanobacteria of some sort already had developed then. That the African fossils of similar age do represent remains of contemporary bacteria has not been challenged.

Another striking feature is the presence of so-called *heterocysts* in some of the fossil cyanobacteria. These are specialized cells that are slightly larger than the neighbouring cells in the filamentous colonies. Heterocysts cannot perform oxygenic photosynthesis, but only cyclic photophosphorylation and their main function is nitrogen fixation, a process that can take place only under anoxic conditions. Their presence among the fossils is also a strong argument in favour of the cyanobacterial nature of the fossils and of contemporary O_2 production.

The earliest confident remains of eukaryotic cells are aged to 1.8 billion years although some 2.1 billion-year-old fossils perhaps also represent remains of eukaryotes. Chemical markers suggest the presence of sterols (characteristic for eukaryotic organisms) in deposits aged no less than 2.7 billion years. As already mentioned, the possibility of leaving recognizable fossils of eukaryotes is generally poorer than is the case for bacteria. The so-called *acritarcs* were equipped with some sort of cell walls or tests (as known from many extant protists) (see Figure 8.4, and also Plate 3). Most were probably planktonic protozoa or protophytes and it is likely that they represent many different lineages of protists. They are especially well known from the Grand Canyon and from Jixian province in China, but they have also been found in many other places. The oldest arcritarcs measured only 60–200 μm, but younger deposits include much larger forms that disappeared again in the Phanerozoic. It is tempting to assume that these larger forms were out-competed by multicellular animals.

Extant stromatolitic microbial mats

Modern stromatolitic mats should be discussed in the context of Precambrian life because they represent a convincing analogy to the fossil mats. Thin cyanobacterial mats are very common, for example in shallow water, on tidal flats, and on dead corals, and they thrive wherever recurrent desiccation or anoxic episodes discourage animals. These mats are usually ephemeral because they are destroyed by storms or because they are periodically colonized by animals. Modern stromatolitic cyanobacterial mats are known from places with extremely high salinities such as certain tropical or subtropical lagoons and lakes and in evaporation ponds used for salt production. They also occur in geothermal springs (Figure 13.4; see also Plate 5). They can be made artificially in the laboratory by removal of animals from marine sediments, then exposing them to a suitable light intensity. This requires some patience because cyanobacterial mats grow slowly (1–2 mm per year). Most extant mats, except those growing at very high temperatures, usually also contain various phototrophic and heterotrophic unicellular eukaryotes; in this respect they must differ from the oldest (> about two billion years) mats.

Fig. 13.4 (See also Plate 5) (Top) Cyanobacterial mats in a hot spring in Yellowstone Park. (Bottom) The hypersaline lake 'Solar Pond' in the Sinai Desert. The bottom is covered by a stromatolitic cyanobacterial mat roughly one metre thick, the surface of which is responsible for the yellow–green colour. Perhaps the picture provides an impression of shallow water habitats through three billion years of the Earth's history: the land is barren and apparently lifeless while the bottom of shallow water habitats is covered by stromatolitic microbial mats. (M. Kühl)

The biological activity of these mats is primarily based on oxygenic photosynthesis of cyanobacteria and to a considerably smaller extent on anoxygenic photosynthesis a little deeper in the mat. By far the highest biological activity is confined to the upper 2 mm of the mats, due to limited light penetration. This zone consists of a layer of tightly interwoven cyanobacterial filaments; usually there are several species that occur at different depths. Beneath the cyanobacteria, there is a 3–4 mm thick layer with purple, brownish, and green colours that disclose the presence of different kinds of phototrophic sulphur bacteria. Beneath this layer the mats are black from precipitated iron sulphides (Figure 13.5; see also Plate 6). The mats also harbour a variety of other kinds of bacteria that together represent most types of known energy metabolism. Mats are typically anoxic beneath 4–5 mm in the light and beneath 0–1 mm in the dark. Sulphate reduction is the most important process for degradation of organic matter in the anaerobic zone of the mats.

The mats have a jelly-like or rubbery consistency, but are often hardened by layers of deposited carbonate minerals (Figure 13.5; see also Plate 6). Carbonates deposit due to the intense photosynthesis in the surface layers that also renders the water alkaline. The mats thicken slowly, in part due to the deposition of carbonate and in part to the empty mucous sheaths of the cyanobacteria. These sheaths, or at least some of their components, degrade very slowly deeper in the mat. The carbonate

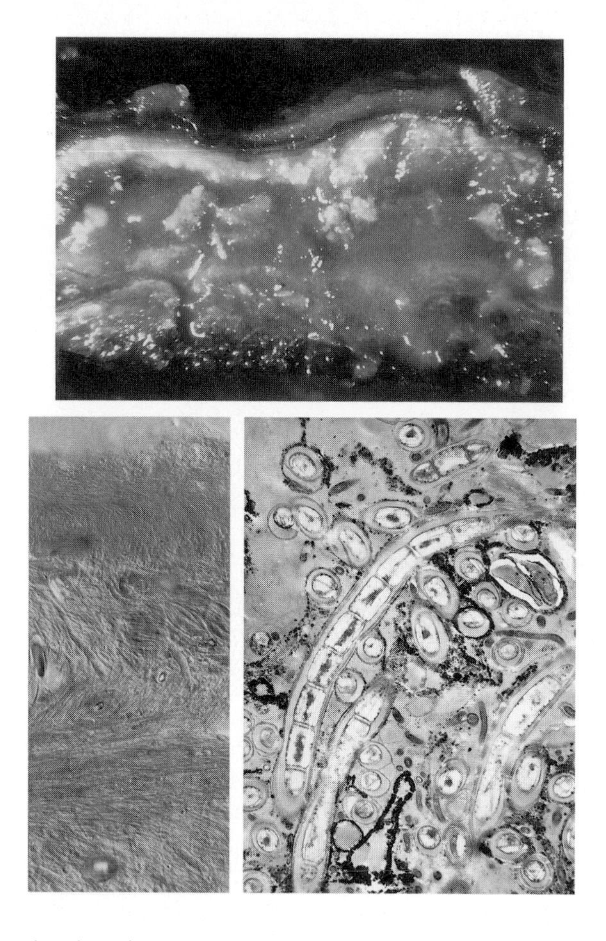

Fig. 13.5 (See also Plate 6) A section through an artificial stromatolitic cyanobacterial mat grown in the laboratory in the absence of animals. It is about three years old and 7 mm thick. Yellowish and dark green layers with cyanobacteria are seen in the upper 1 mm above a layer with carbonate precipitation. The red, brownish, and greenish colours deeper in the mat are due to different types of bacteria with anoxygenic photosynthesis. (Bottom, left) A freeze-section of the upper 2 mm of the mat with filaments of the cyanobacteria *Calothrix*, *Pseudanabaena*, and *Phormidium*. (Bottom, right) Electron micrograph about 1 mm beneath the mat surface including *Pseudanabaena* filaments and various other bacteria. (Originals)

deposition is, of course, the reason why the fossilized mats have been preserved as carbonate rock. Metre-thick extant mats are known (e.g. from Solar Pond in the Sinai Desert, Figure 13.4 and Plate 5) and they must be at least 1000 years old.

A final aspect of these mats should be mentioned. During most of the Precambrian the atmosphere had a low oxygen tension: about two billion years ago it was only 1–2% of the current value (see the following section). Nevertheless, very high oxygen concentrations must have occurred locally at the surface of cyanobacterial mats. Figure 13.6 shows the depth distribution of oxygen tensions in a modern cyanobacterial mat in the light and in the dark. In the light, the oxygen tension may reach up to four times atmospheric saturation and so bubbles of pure oxygen form. Something similar must have occurred in the oldest Precambrian mats; high local oxygen tensions would have been produced even if the atmospheric oxygen tension at the time was close to zero.

The rise of atmospheric oxygen

The origin of oxygenic photosynthesis did not signal a rapid increase in atmospheric O_2 to the present value of about 21% (Figure 13.2). Initially, oxygen would be quickly consumed through oxidation of reducing compounds such as ferrous iron and sulphide that initially were stable in an anaerobic atmosphere. This process took a long time and to some extent it continues today in that reducing minerals are oxidized when exposed to the atmosphere. Also, it must be assumed that aerobic respiration developed immediately after the advent of oxygenic photosynthesis so that organic matter produced by photosynthesis became re-oxidized, consuming an amount of oxygen roughly equivalent to the amount produced. The fact that there was an excess of oxygen at all in terms of accumulation in the atmosphere and in the oxidation of iron and sulphur, therefore depended on the burial of a corresponding amount of organic matter (or iron sulphides of biological origin) in sediments. The free oxygen that is present on the Earth + oxidized Fe and S must therefore always be balanced by the amount of fossil organic matter + biologically-produced sulphide and methane.

The so-called *banded iron formations* (Figure 13.7; see also Plate 7) represent the earliest signs of the presence of small amounts of atmospheric oxygen. Banded iron formations may be hundreds of metres thick; they are known from the area around Lake Superior, from Greenland, Western Australia, South Africa, Russia, India, and Minas Gerais in Brazil and they are exploited as iron ores. The oldest are 3.8 billion years from the Isua formation in Greenland and the youngest are about 1.9 billion years. They were also deposited during a shorter period, about 750 million years ago. Banded iron formations consist of horizontal layers of more or less completely oxidized iron: that is, different mixtures of Fe_2O_3 and FeO with a mean composition of Fe_3O_4 (magnetite) or about equal amounts of Fe^{2+} and Fe^{3+}. Banded iron formations also often include horizontal layers of quartz. The horizontal banding may reflect seasonal variation in the conditions for deposition.

The general interpretation of these formations is that the seas were for a long time basically anaerobic and under these conditions seawater could contain large amounts of dissolved Fe^{2+} (in contrast with the almost insoluble Fe^{3+}). As the atmosphere began to contain low levels of O_2, ferric iron could precipitate in the surface layer

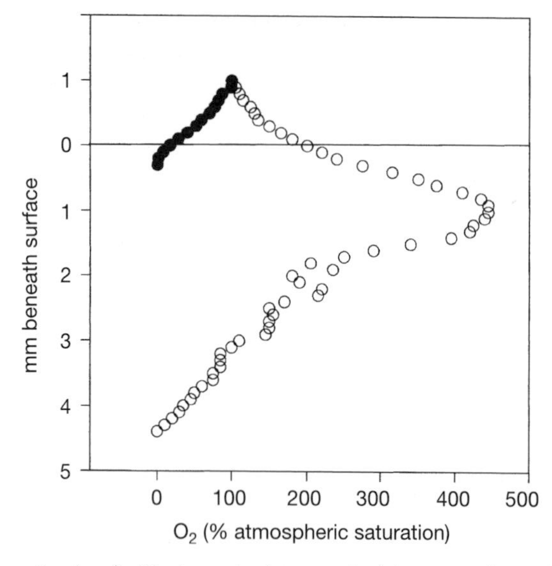

Fig. 13.6 Oxygen tension (in % atmospheric saturation) in a cyanobacterial mat in the light and in the dark.

Fig. 13.7 (See also Plate 7) Banded iron formation in Western Australia. (D. E. Canfield)

and sediment to the bottom and thus create the banded iron formations. It is also plausible that the oxygen for iron oxidation came from *in situ* photosynthesis of planktonic cyanobacteria in the surface water. About 1.9 billion years ago banded iron formations were replaced by *red-beds*; these are alluvial deposits of fully oxidized iron. At that time the oxygen content had reached a much higher level, the surface of the Earth had become rusty, and the largest buffer against accumulation of atmospheric oxygen had been saturated.

It should be mentioned that there are alternative explanations for the banded iron formations. One possibility is that the oxidation was anaerobic and caused by photosynthetic bacteria that use Fe^{2+} as a reductant in photosynthesis (this process is also known from some extant purple bacteria). Since banded iron formations usually contain little organic matter this may not be a likely explanation. It has also been suggested that the iron oxidation took place anaerobically through photochemical reactions, because the process:

$$Fe^{2+} + H^+ \rightarrow Fe^{3+} + H_2$$

may take place with exposure to short wavelength light. These processes may well have contributed and it is also possible that different mechanisms were of different importance during the enormous time span of banded iron formation deposition.

There is other, more direct evidence that the atmospheric O_2 content slowly increased during the Precambrian and that it had reached a level of about 1–2% of the present level about two billion year ago. Pyrite (FeS_2) and uraninite (UO_2) are, in the absence of O_2, insoluble and stable minerals. Grains of these minerals have been found in >two billion-year-old palaeosols (terrestrial deposits). This would be impossible if the atmosphere had contained more than 1–2% of the present oxygen tension. Pyrite would then quickly oxidize (to sulphate + rust) and uraninite would oxidize to water soluble uranium salts. Just as the banded iron formations are commercially important, the commercially important uranium deposits formed for similar reasons about two billion years ago. After the atmospheric O_2 concentration had become sufficiently high soluble uranium salts formed through oxidation on land and were washed out into the sea. The continental seas were then still predominantly anoxic and reducing and so uranium precipitated and accumulated in an insoluble form again.

It is therefore possible to establish an atmospheric O_2 tension that was 1–2% of the present level about two billion years ago. It can also be established that about 600 million years ago the O_2 tension must have reached a level that was at least 10% of the present level. This is because macroscopic animals that arose at that time require a minimum oxygen tension of about 10% for survival (see also Chapter 14).

It is probable that only the surface waters of oceans and of continental seas were aerobic during the Precambrian. The seas must in this respect have been like the Black Sea today, that is, permanently anoxic beneath 100–200 metres depth.

We do not know the development of atmospheric O_2 in any detail for the Phanerozoic, but atmospheric O_2 levels have probably not changed much since Cambrian. It has been suggested that O_2 levels were higher during the Carboniferous (this should explain the presence of giant dragon flies and it is also consistent with the large deposition of organic matter that eventually became coal). There have been several episodes when the deeper parts of continental seas (and presumably of the oceans) were anoxic and sulphidic, for example during the Cretaceous. This is reflected in the deposition of black shales over large areas. The reason is probably not fluctuations in atmospheric O_2, but rather the course of ocean currents. Today, oceanic surface water sinks down in the Arctic and Antarctic to aerate the deep sea. In geological periods with warmer climates this 'conveyor belt' transport of aerated surface waters to the deeper parts of the sea was absent resulting in widespread anoxia at greater depths.

A sort of mass balance for atmospheric oxygen (including that dissolved in the sea) is shown in Figure 13.8. The excess of oxygen depends on a corresponding accumulation of organic matter (plus biologically reduced sulphur and methane) that has become buried in sedimentary deposits. About 58% and 38% of this oxygen has been used for the oxidation of iron and sulphur, respectively during the 3.5–3.8 billion years of oxygenic photosynthesis. The remaining 4% (about 1.2×10^{15} tonnes) is found as free O_2. The turnover time of this atmospheric oxygen is only about 4000 years and is caused by oxygenic photosynthesis and respiration—processes that almost balance so that the biological turnover is vastly faster than the slow accumulation of organic matter. Given the crude estimates in Figure 13.8, there is altogether an excess of 3×10^{16} tonnes of oxygen (of which the greater part is in the form of oxidized iron and sulphur and only 4% as O_2). Over roughly four billion years this corresponds to a net accumulation of about 7.7×10^6 tonnes per year. The annual turnover due to photosynthesis and respiration corresponds to $(1.2 \times 10^{15})/4000 = 3 \times 10^{10}$ tonnes O_2 per year; that is production and respiration is about 20 000 times faster than the slow accumulation of organic matter in sedimentary rocks. This calculation, of course, assumes that both the production rate and the rate of accumulation of un-mineralized organic matter have been constant over geological time. Some evidence indicates that this has been the case since roughly the second half of the Precambrian.

Finally it may be asked whether the oxygen contents of the atmosphere will continue to increase. It is generally assumed that atmospheric O_2 today (and during the last 400–500 million years) is controlled by various mechanisms that maintain it at a relatively constant level. An atmospheric O_2 level that is somewhat higher than

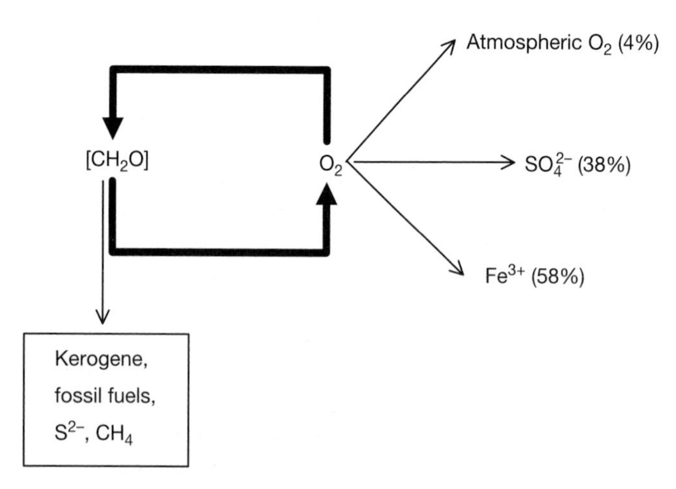

Fig. 13.8 Approximate fate of the net production of oxygen that has been produced since the origin of oxygenic photosynthesis and corresponding stoichiometry to the amount of fossil organic matter plus sulphides and methane of biological origin.

the present one will lead to self-ignition of organic matter in the air. It is therefore possible that the level of spontaneous forest fires has controlled atmospheric oxygen tension since the Silurian or Devonian when land was inhabited by life. It is also possible that oxygen toxicity sets an ultimate limit. It has finally been suggested that the availability of phosphate regulates atmospheric oxygen levels indirectly. Under oxic and neutral conditions phosphate tends to precipitate in the form of Ca or Fe phosphates that are unavailable to plants, algae, and bacteria. Under acid and chemically reducing conditions phosphates are again released in a soluble form. An increased oxygen level will expand oxic habitats, thus decreasing the availability of dissolved phosphate and consequently limiting primary production. In this way, oxygenic photosynthesis would be self-regulating. All this exemplifies the large, and perhaps un-surmountable difficulties in obtaining a quantitative understanding of global biogeochemistry, such as the precise composition of the atmosphere. Today we probably have a rather complete qualitative understanding of the major biotic and abiotic processes, but their interactions are so complex that it has not, for example, been possible to explain why the atmosphere now contains exactly 21% O_2.

The formation of an atmospheric ozone layer was also a consequence of atmospheric O_2. This was a prerequisite for the colonization of land since the ozone layer would provide protection against the intense UV radiation from the sun. It has been estimated that an atmospheric O_2 level of about 10% of the present value would be sufficient to achieve this, and this level was certainly reached when plants and animals colonized land some 400 million years ago. On the other hand, it is a fact that microbial life must have thrived in quite shallow water long before the atmosphere had reached such O_2 levels. These organisms must have possessed more developed versions of the mechanisms (mainly carotenoids) that today protect phototrophic organisms (such as cyanobacteria) against the adverse effects of short wavelength light.

The development of biogeochemical cycling

Some scientists have suggested that the first life was thermophilic and lived at high temperatures in hot springs. In support of this view is the fact that an early branch among the eubacteria (*Thermotoga*) is thermophilic and so are the thermophilic archaebacteria that are also close to the root of the universal tree (Chapter 12 and Figure 12.4). This is coupled to the idea that life arose in connection with geothermal phenomena such as hot springs. Association with such geothermal habitats could explain the access to necessary reducing compounds (e.g. H_2, H_2S) before oxygenic photosynthesis could produce copious amounts of reduced organic compounds. Geothermal heat could also have provided polyphosphates from phosphate minerals. So this is a possibility, but molecular data do not make this a necessary conclusion. Several authors have proposed deep sea hydrothermal vents as a site for the origin of life. But this somewhat romantic idea has little appeal. Deep sea hydrothermal vents provide substrates for some anaerobic processes such as H_2 oxidation with CO_2 or SO_4^{2-} which are realized in these habitats, as they are in many other habitats. But the rich life discovered around hydrothermal vents, and the splendid underwater photographs, that have inspired ideas about 'the cradle of life' in such habitats, is based not only on sulphide or methane seeping up from the ocean floor, but also on the presence of oxygen in the water—and this oxygen is derived from oxygenic photosynthesis in the photic zone of the biosphere. And when life arose oxygen was absent! Presumably light energy was a necessary pre-condition for the origin of organic molecules and of life, but even if the origin of life depended on only purely chemical energy sources, there is no reason to relegate the 'origin' to the deep sea. The origin of the hydrothermal vent biota we observe today is surely relatively recent—also because the ocean bottoms have probably been anoxic for periods during the Phanerozoic.

It is probable that the earliest biogeochemical cycling was based on the transformations of iron or sulphur or both. Figure 13.9 shows such possibilities for biologically-mediated iron and sulphur cycles. Their realism as a model for early microbial ecosystems is supported by the existence of purple and green phototrophic bacteria that use Fe^{2+} and H_2S as electron donors in photosynthetic processes, as well as several types of eubacteria and archaebacteria that use Fe^{3+} and SO_4^{2-} as electron acceptors in respiration. The two cycles could well have been coupled—also because there are extant bacteria that make a living from the oxidation of sulphide with ferric iron. The two opposing processes, that is photosynthesis driven by light energy, and respiration driven by downhill chemical processes, use the same enzymatic apparatus (Chapter 7) and may both take place within phototrophic bacteria in the light and in the dark respectively. It is, however, likely that forms

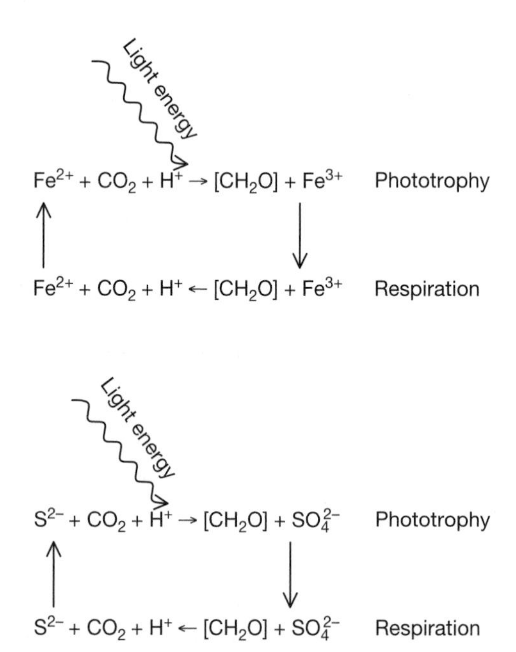

Fig. 13.9 Microbial iron and sulphur cycles that may have dominated biogeochemical cycling before the origin of oxygenic photosynthesis and aerobic respiration.

that had lost the ability of phototrophy quickly arose so that the cycles were driven by specialists in photosynthesis and respiration, respectively.

The early anaerobic seas must have contained large amounts of dissolved Fe^{2+}, and sulphide was then largely bound as ferrous sulphides. The large scale availability of sulphur therefore has become a reality after oxygen began to oxidize iron. We can therefore speculate about an early 'iron world' where microbial energy metabolism depended on the redox couple $Fe^{3+} \leftrightarrow Fe^{2+}$ followed by a 'sulphur world' depending on the redox couple $SO_4^{2-} \leftrightarrow S^{2-}$, and that eventually the redox couple $H_2O \leftrightarrow O_2$ dominated. It is therefore possible that a quantitatively significant role of the sulphur cycle was conditioned by the origin of oxygenic photosynthesis, so that most iron had become oxidized. But it is still reasonable to assume that processes depending on the oxidation and reduction of sulphur compounds played a great role in a world with a very low atmospheric oxygen content.

When we reach the period following the origin of oxygenic photosynthesis—which actually represents almost the entire period for which we have any geological evidence—we are on firmer ground. This is because we can assume that all the types of microbial energy metabolism that we know today were then already established. Quantitatively, there have probably been great changes, in particular during the early Precambrian, but qualitatively it is unlikely that much has changed.

Figure 13.10 represents a kind of general biosphere model. It shows that oxygenic photosynthesis produces a chemical potential: $O_2/[CH_2O]$. Here $[CH_2O]$ represents organic matter in which the oxidation level of C is like that in glucose. The vertical axis to the right shows that the oxidation of $[CH_2O]$ with O_2 releases about 120 kJ per mole transferred electrons. This is the most energy efficient biological process known, although oxidation of H_2 or CH_4 by oxygen is nearly as profitable. All three types of oxidation are used in the energy metabolism of many kinds of organisms, as well as for heating houses or driving rockets to the moon. (It should be noted that the changes in free energy shown in Figure 13.10 apply to standard conditions that do not necessarily apply under physiological conditions. Also the conserved energy available to organisms in terms of ATP may be one-half or less of the stated values. The ranking of the processes, however, also applies to the energy metabolism of real organisms.) The central vertical axis represents standard redox potentials for the redox couples. In order to be thermodynamically possible the oxidizing redox couple (to the left) must have a higher redox potential than that of the oxidized redox couple (values to the right).

The reason why oxidation processes that involve electron acceptors other than oxygen are used today is that anaerobic habitats are still widespread. Oxygen has a low solubility in water and is often used faster than it can be supplied from the environment through molecular diffusion. When oxygen is depleted, *denitrification* (nitrate respiration) is the process of choice—it is apparently almost as efficient as oxygen respiration. Denitrification must have evolved after the origin of oxygenic photosynthesis since nitrate could not have been present on Earth's surface in the absence of O_2. Even today, nitrate is scarce in most habitats, and while denitrification is important in other contexts of biogeochemistry it plays a modest role in the oxidation of organic matter. When nitrate has become depleted in

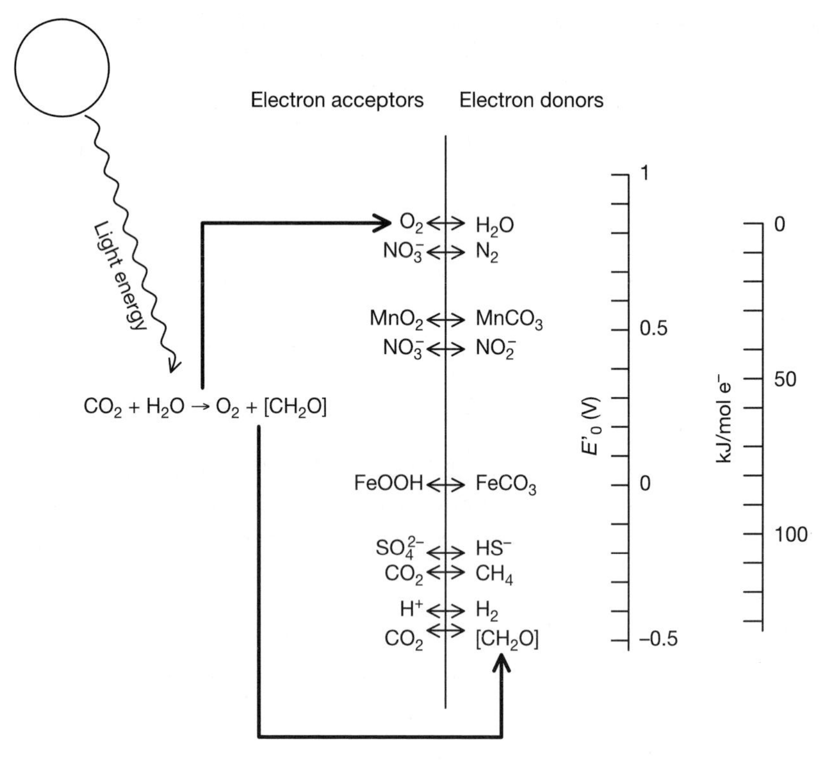

Fig. 13.10 A simplified biosphere model. Oxygenic photosynthesis provides a chemical potential that is used by other organisms through coupled redox reactions. For further explanation, see text. (Redrawn from Fenchel and Finlay, 1995.)

anaerobic habitats the electron acceptors are used in the sequence: Mn^{4+}, Fe^{3+}, SO_4^{2-}, and CO_2.

A completely anaerobic ecosystem could be based on anoxygenic photosynthesis (using, for example, sulphide as electron donor leading to sulphate and organic matter) and sulphate respiration (based on the oxidation of organic matter with sulphate as oxidant). Such ecosystems are still realized in certain special habitats. It is seen from Figure 13.10 that this is a 'low energy system': oxidation of H_2 or organic matter with sulphate appears to generate only about one-fifth of that resulting from oxidation with O_2 (in fact it is even less because of the high activation energy of the sulphate molecule). Even long after oxygenic photosynthesis had arisen, such anaerobic systems would have been substantially more important in the biosphere than they are today, because the O_2 tension of the atmosphere remained relatively low through most of the Precambrian. Nevertheless, the origin of aerobic respiration, with its high energy yield, was the prerequisite for the evolution of all higher life.

Somewhat less generalized presentations of the three basic biologically driven element cycles are shown in Figures 13.11, 13.12, and 13.13. They are all simplified in that some intermediate oxidation levels have been omitted. The cycles describe the most important processes in the biosphere today and presumably also three to four billion years ago. Cycling of other elements could also be described in this way, but they are generally simpler and quantitatively less important in the biosphere. The three elements (C, N, S) resemble each other insofar as they are important constituents of all organisms and at the same time they are involved in energy-generating redox processes in cells. Transformations of these elements affect biological and non-biological cycling of other elements as well.

The single most important process in the carbon cycle (Figure 13.11) is the assimilatory reduction of CO_2 to organic matter through photosynthesis (and to a much smaller degree through chemoautotrophic processes). The organic matter is then re-oxidized, often through complex intermediate steps via fermentation and respiration with different electron acceptors and eventually with O_2 so that CO_2 is regenerated. The low atmospheric concentration of CO_2 means that the mean residence time is only about seven years, and this explains why the burning of fossil fuels during the last century has resulted in a measurable increase in atmospheric CO_2. A small fraction of the produced organic matter is fossilized in sediments and some CO_2 is deposited as carbonate minerals. Some of this CO_2 is recovered through volcanic out-gassing, but the net accumulation of organic matter and carbonates has resulted in a slow decrease in atmospheric CO_2 concentration over geological time. Ultimately (and notwithstanding the current concern for the opposite process) the eventual disappearance of atmospheric CO_2 may mean the end of life on Earth because photosynthesis is not possible below a certain atmospheric CO_2 tension.

The sulphur cycle (Figure 13.12) differs from the carbon cycle in several respects. Sulphur occurs in the most reduced form (as sulphide) in living cells. Many bacteria, algae, and plants can utilize sulphate as a source of sulphur through assimilative reduction (not shown in the figure). Degradation of organic matter releases sulphide which, however, is rapidly oxidized by various colourless sulphur bacteria. In anaerobic habitats (and especially in marine sediments) sulphate reduction is the most important respiratory process. The resulting sulphide is then re-oxidized by colourless sulphur bacteria in the presence of oxygen or nitrate or anaerobically by photosynthetic sulphur bacteria. Sulphur is lost to the biosphere as pyrite in sedimentary rocks, but it is recovered through geothermal processes or biological oxidation where pyrite is exposed to atmospheric oxygen. Sulphate was probably not present on the surface of the primordial Earth, but was produced by photosynthetic bacteria and biologically mediated oxidation by O_2. The enormous reservoir of sulphate in seawater (the SO_4^{2-} concentration in seawater is about 28 mM) and in anhydrite (gypsum) deposits derived from evaporation of seawater is thus the result of biological activity over billions of years.

The primordial atmosphere must have contained N_2, but it probably also contained the most reduced form of nitrogen as NH_3 that would have been necessary for the non-biological production of amino acids and other N-containing organic compounds. More oxidizing forms of N were probably not present before the atmosphere contained O_2. For reasons that are not clear, organisms that use ammonia as an electron donor in a photosynthetic process have apparently never evolved and so there was no way in which oxidized forms of N could occur prior to free O_2.

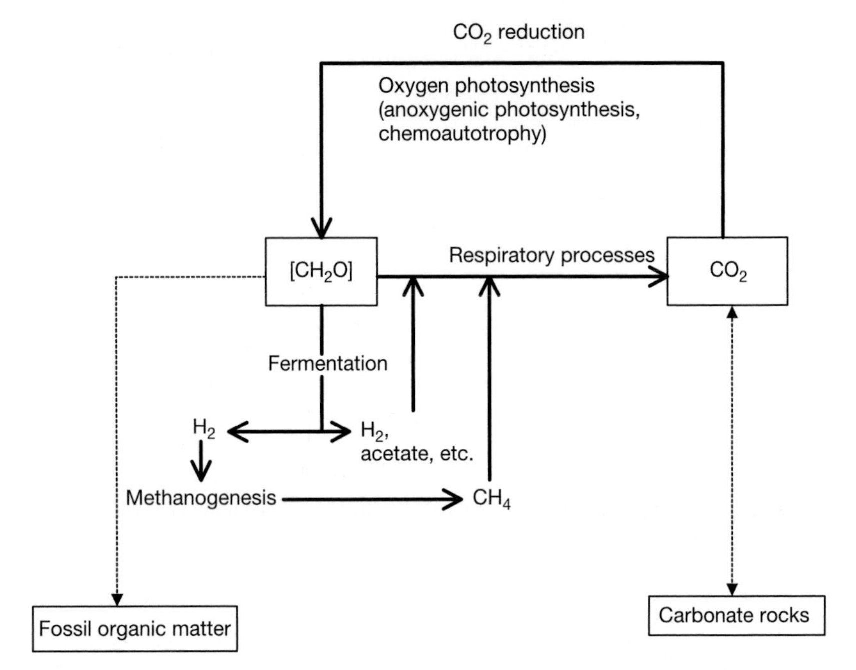

Fig. 13.11 The biogeochemical carbon cycle. For further explanation, see text.

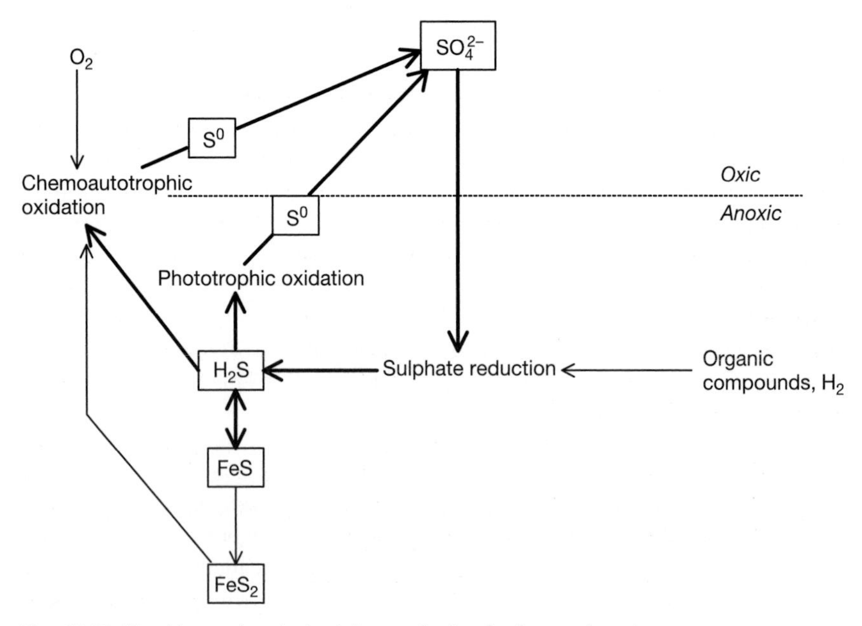

Fig. 13.12 The biogeochemical sulphur cycle. For further explanation, see text.

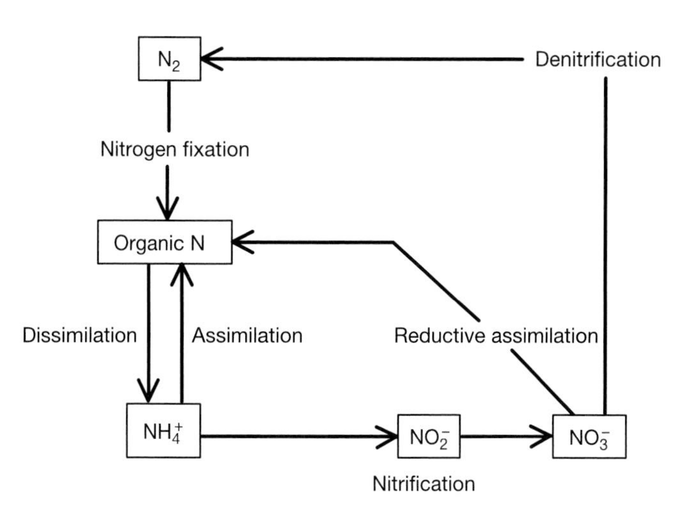

Fig. 13.13 The biogeochemical nitrogen cycle. For further explanation, see text.

When oxygen appeared, all this changed. An N_2/O_2 mixture is thermodynamically unstable and should lead to nitrogen oxides. However, the activation energy of the N≡N bond is so high that spontaneous oxidation does not generally occur. During electric discharges (lightning), oxidation of N_2 does occur and over longer time periods much of the atmospheric nitrogen should have ended up as dissolved nitrogen oxides in the sea. Furthermore, biologically-mediated nitrate production takes place in an indirect way (Figure 13.13). Many bacteria are capable of fixing N_2. This is an energy-requiring process and so it occurs only if other combined forms of nitrogen are unavailable. Through nitrogen fixation N_2 is reduced to NH_4^+, that is the form occurring in organic compounds. The ammonia is again released during the mineralization of organic matter. *Nitrifying bacteria* then oxidize the ammonia no nitrate in the presence of oxygen. So, together with non-biological oxidation of atmospheric N_2, this would result in an accumulation of nitrate were it not for the existence of *denitrifying bacteria.* These organisms use nitrate as an electron acceptor in respiration and the end-product of this process is N_2. This is the only way in which the N_2 content of the atmosphere is replenished. The residence time of atmospheric N_2 is long, however (about 50 000 years). The nitrogen cycle became complex after the advent of free oxygen and N_2 was no longer a conservative constituent of the atmosphere.

It is characteristic of the major biogeochemical cycles that all the processes can be catalysed by bacteria and they were all probably established shortly after the advent of oxygenic photosynthesis and have probably not undergone fundamental qualitative change since then. Even today, many biogeochemical key processes are exclusively carried out by bacteria; such processes include the production and oxidation of methane, sulphate respiration, anaerobic (phototrophic) sulphide oxidation, nitrogen fixation, nitrification, and denitrification.

Precambrian glaciations

Traces of the oldest known ice-age are dated to about two billion years ago. Ice-ages can be recognized from moraines (those from the most recent ice-age are well-known features of Northern European landscapes, in parts of North America, and elsewhere in mountainous regions). Abrasion marks on cliffs formed by rocks transported in overlying glaciers is another sign of past ice-ages. Melting icebergs leave their marks by dropping stones into otherwise soft, fine-grained sediments, and these stones can later be found in sedimentary rocks.

Although the glaciers retreated about 10 000 years ago, we do live in a relatively cold period of Earth history in comparison with the average Phanerozoic climate. In addition to the most recent (Pleistocene) ice-age period (that has so far lasted about 1.5 million years punctuated by several shorter inter-glacial periods with milder climates) there were two earlier periods with glaciations during the Phanerozoic. An apparently more intense cold period occurred during the late Precambrian in the period from about 900 to about 590 million years ago. Altogether this period shows signs of four glaciation events. In particular, the two youngest: the Sturtian (720 million years) and the Varangian (610–590 million years) are considered especially intense and long-lasting with glaciers developing at both poles.

Ice-ages tend to be self-propagating. When a period of snow-cover is prolonged during spring and occurs at lower latitudes, this increases Earth's *albedo* (the amount of light that is reflected from the Earth's surface rather than being absorbed as heat) and so temperatures fall. Conversely, if the climate becomes milder, the albedo decreases, further accelerating warming.

There are many tentative hypotheses for why periods of repeating ice-ages come and go. These include cyclic variations in the Earth's trajectory around the sun and the tilt of Earth's axis, the configuration of the continents of the Earth's surface affecting ocean currents, and intense mountain building as a nucleation factor for glacier formation. With respect to the late Precambrian glaciations, it has been suggested that the cause was a fall in atmospheric CO_2 due to intense photosynthetic activity—a sort of inverse greenhouse effect. The reason why the Earth did not end up as a permanently ice-covered planet should then be that glaciations reduced photosynthesis while volcanic out-gassing replenished the atmospheric CO_2 content so that a warming of global climate could take place. And so a cyclic pattern of ice-ages and warm periods would result. How the Earth eventually escaped this vicious circle until the next ice-age at the Ordovician–Silurian transition has not been explained, however.

Some geologists believe that the late Precambrian glaciations were the most dramatic ones in the Earth's history and it was, for periods, almost totally ice covered—the term 'Snowball Earth' has been used to describe this. Other geologists disagree, and think that at least low latitude waters were always open. Actually, while there is strong evidence of the existence of these late Precambrian ice-ages, there is no compelling evidence indicating exactly how intense and widespread they were.

There have also been suggestions that these events had profound evolutionary effects including mass extinctions and the rise of animals. It is true that the first appearance of animals occurs shortly (in a geological time-scale) after these glaciations. But how, and to what extent, these glaciations had any profound effect on evolution is so far only a matter of speculation.

The Gaia hypothesis as pseudo-science

It is beyond doubt that biological processes have profoundly affected the chemical and physical environment on the surface of Earth. Inspired by this, the British chemist J. E. Lovelock has developed what has been called the 'Gaia hypothesis' (*Gaia* is a kind of Mother Earth Goddess in Greek mythology). It implies that all living organisms, together with the atmosphere and the oceans, constitute some sort of super-organism that controls the chemical environment to its own ends. Thus 'Gaia' controls temperature, the reduction–oxidation potential, pH, etc., in a way that creates optimal conditions for life. A cited example is that biological processes caused a reduction in atmospheric CO_2 as the radiation from the sun increased—and this may be regarded as a kind of Gaia strategy to maintain a constant and tolerable temperature regime on the Earth's surface. This is presumably what actually happened, but it was hardly due a strategy implemented by Gaia.

To be kind it could be said that the Gaia hypothesis is a sort of metaphor. More to the point it is pseudo-science devoid of explanatory power. As Gaia has enjoyed a certain popularity bordering on the status as ersatz religion, a few comments are appropriate in the context of this book. The founders of the Gaia hypothesis have more recently retracted some of the more extreme teleological aspects—but if this process is carried to its conclusion, then Gaia will simply dissolve into a mixture of trivialities and more traditional science.

It is a common observation that biological as well as non-biological systems have cybernetic properties in the form of negative feedback. In an ecological context, for example, death rates increase and birth rates decrease as a species population grows, due to intraspecific competition, and this often leads to relatively stable population sizes. Similarly, populations that are mutually interdependent, such as populations of photosynthetic and heterotrophic species, tend to stabilize each other. Non-biological phenomena may have similar properties: a warming climate leads to a denser cloud cover that decreases the Earth's albedo resulting in cooling. But none of this implies regulation by a super-organism.

Lovelock has stressed that the fact that the physical–chemical environment is 'optimal for life' is a sort of evidence for the Gaia hypothesis. But this is just a posteriori recognition of the fact that there is still life on Earth—and if this were not the case, then there would be no one to write about it. The reason why conditions appear optimal for life is only that organisms, through Darwinian evolution, have adapted to the prevailing chemical and physical environment—whether changes were caused by vagaries of geological or climatic events or through the activities of other organisms. The slow or fast, but under all circumstances dramatic, changes that have happened over geological time (oxygen in the atmosphere, climatic fluctuations, impacts of comet or asteroids) have undoubtedly resulted in 'set backs' for life and extensive extinctions of evolutionary lineages. Palaeontology has recorded several episodes of mass extinctions during the Phanerozoic period. But some organisms have always survived and adapted to new conditions through evolutionary processes. Finally it is quite unclear what is really meant by 'optimal conditions for life': polar bears may approve of a change in climate that will be regretted by giraffes!

There are innumerable examples of organisms that have adapted to the presence of other organisms, and sometimes such relations have evolved into symbiotic or mutualistic consortia. All this can be explained through Darwinian selection mechanisms. But there is no mechanism that could explain how 'life' should collectively adapt or in some teleological way regulate the physical and chemical environment on the surface of the Earth. Gaia cannot be a unit for natural selection—and 'she' cannot provide any real explanation for why the biosphere developed the way it did.

Chapter 14

Transitions during the evolution of life

Contemporary understanding of evolution (often referred to as the *Neo-Darwinian Synthesis*) explains directional evolution as the replacement of alleles in species populations driven by natural selection. An overwhelming amount of theoretical and experimental documentation as well as data deriving from palaeontology and natural history shows that this description is fundamentally correct. It is not, however, always appreciated that the effect of natural selection is often the maintenance of stable phenotypes over long periods of geological time. Many textbooks on evolution show a picture of the brachiopod *Lingula* that—at least as far as external morphology is concerned—has remained totally unchanged from the Ordovician until today (brachiopods are a kind of marine invertebrates resembling bivalves). The phenomenon is called *stasis*. Mutations will occur continuously. Some of these are neutral and may become fixed through genetic drift. But to the extent that the environment remains stable, then phenotypes will achieve a kind of local fitness maximum. Thereafter, selection will act to remove any mutant genes that result in deviations from this locally optimal phenotype.

Occasionally this pattern is punctuated during (in a geological time-scale) short periods by rapid evolution of new phenotypes, so-called adaptive radiations. In a smaller scale this happens when a new 'empty' habitat is colonized by a single or few species. Over a short period there is a high rate of speciation and different lineages adapt to different ecological niches. Examples are the cichlid fishes of the large African lakes and Darwin's finches on the Galapagos Islands. Adaptive radiations also occur when new organizational levels of organisms appear on the evolutionary scene. The most famous example is the Cambrian explosion—the rapid diversification of multicellular animals in the early Cambrian that led to practically all the main types of extant invertebrate groups.

This concluding chapter will consider the three major transitions of this sort during evolution: the origin of the prokaryote cell, the origin of the eukaryotes, and the origin of multicellular organisms. We will especially be concerned with the question of why the two latter events happened so late. It will be recalled that for about the first two billion years (or roughly half the time there has been life on Earth) there were only bacteria. About two billion years ago the eukaryotes appeared, and multicellular organisms arose only some 600 million years ago.

It has already been stressed that we know almost nothing about how the first real cell came about. But we know, combining evidence from palaeontology and molecular trees, that 'modern' bacteria must have existed a short time after the origin of life and that most of the main lineages of prokaryotes also diverged very early. This event could suitably be referred to as the first Precambrian explosion. Bacteria have, since, then shown a remarkable degree of stasis.

One reason for this is undoubtedly that the main habitats for free-living bacteria have existed throughout life's history or at least since some oxygen was produced very early by oxygenic photosynthesis. It is also clear that the bacterial level of organization constrains further evolutionary innovation. We have seen that the genome length of bacteria is limited by replication speed. The cell wall and the lack of a cytoskeleton constrain cell size and do not permit uptake of particulate matter (so a predatory habit could not evolve). Basically prokaryotes have little scope to increase complexity and the bacterial phenotype has by and large been preserved from the early Precambrian to this day.

It is likely that primitive eukaryotes also arose very early as the first phagotrophs on the Earth. However, both the molecular phylogenetic tree for the eukaryotes and palaeontology suggest a later adaptive radiation that maybe took place 1.5–2 billion years ago. Whether the term 'second Precambrian explosion' is quite appropriate is not really known: the radiation of the main eukaryote lineages may have taken place over a longer time span. The subsequent radiation of animals in the early Cambrian did, however, take place over a short geological time span, although similar later 'explosions' took place during the colonization of terrestrial habitats by vascular plants, terrestrial arthropods, and vertebrates.

Several authors have previously suggested that the Cambrian explosion was caused by the rise in atmospheric oxygen. This point of view will be elaborated here and extended to the radiation of eukaryotic protists. It will be argued that the critical levels for the atmospheric O_2 tensions that allowed for the radiation and diversification protists and of multicellular organisms where about 1% and 10%, respectively of the present atmospheric level.

There are two key observations in this context. First of all, the primary innovation of eukaryotes was the development of phagocytosis: they were the first predators on the Earth. The earliest forms presumably fed on bacteria and later, after they had diversified, they also ate each other. The ability of phagocytosis was also a prerequisite for the establishment of the endosymbiotic bacteria that eventually became mitochondria and transformed the eukaryotes into aerobic organisms. The second observation is that anaerobic metabolic processes provide a low energy yield as compared with aerobic respiration. The efficiency of energy metabolism determines the growth yield of organisms. We can, as an example, compare a fermenting and an aerobic bacterium, both using glucose as substrate. The former has an energy yield of about 3 moles of ATP per mole of dissimilated glucose. It can be calculated (and experimentally verified) that this leads to a growth efficiency of about 10%, meaning that the utilization 100 g of glucose will result in a bacterial biomass of 10 g. In the case of the aerobe it can produce 32 moles of ATP per mole of dissimilated glucose and this can support a growth efficiency of about 60% (for phagotrophs it is in practise somewhat less: 40% is more realistic).

Ecologists speak about food chains. An example could be unicellular algae that are eaten by copepods. Herrings eat the copepods, mackerels eat herring, and finally, tunas eat the mackerels. Each such step is referred to as a *trophic level*. In practice such linear food chains are rare, and the term 'food web' is more appropriate, insofar as a given species may be represented at several trophic levels—just as we eat herring, mackerel, and tuna—but this does not affect the following considerations. It is obvious that food chains must have a finite length. This is because at each level some of the substrate or food is used for energy metabolism; and phagotrophs digestion efficiencies are usually less than 100%. The energy input into food chains is finite—in the above example it is the photosynthetic rate of the algae and this is fundamentally limited by light or mineral nutrients. Ecology texts usually claim that the length limits of food chains are reached when 1% of the original energy input is left. This is a crude estimate, but we stick to it in the following argument.

Figure 14.1 shows two food chains: an anaerobic and an aerobic one. The aerobic one is identical to the above mentioned plankton food chain. It is assumed that the growth efficiencies are 40% at each level and the graph shows how much of the initial input (arbitrarily assumed to be unity) remains at each trophic level. We see that it allows for six levels before only 1% of the energy input remains. Since the prey: predator length ratio is roughly 1:10 we can also see the food chain allows for very large and complex creatures such as tuna. Turning to the anaerobic food chain we assume an input of dead organic matter. Anaerobic bacteria degrade this material and the bacterial biomass production will be 10% of the input. Anaerobic protozoa eat the bacteria and the biomass production of the protozoa will represent only 1% of the available energy in this ecological system. We see that anaerobic ecological systems allows for only very short food chains. Phagotrophic anaerobic protozoa also

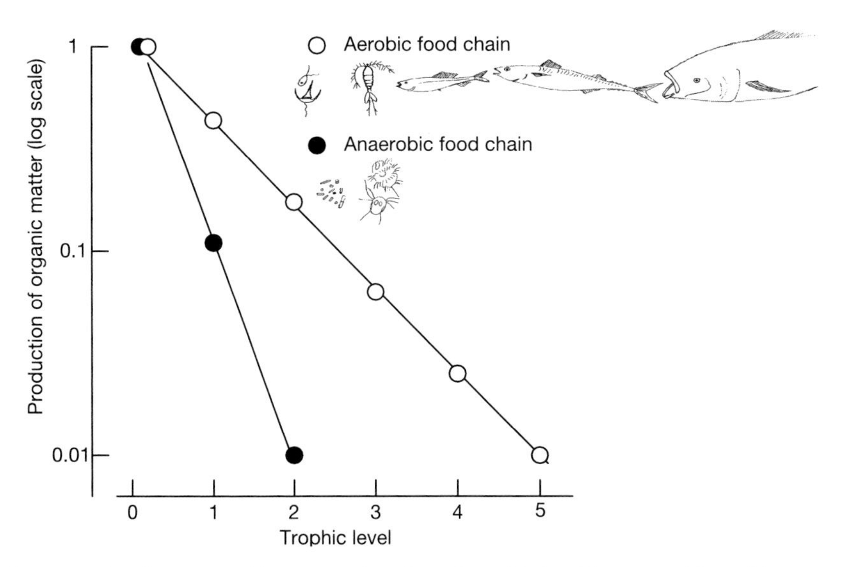

Fig. 14.1 The decrease in biomass as function of trophic level in an aerobic and in an anaerobic food chain. (Redrawn from Fenchel and Finlay, 1995.)

exist today, but as predicted, their diversity is low and they represent only one trophic level in food chains. Long and complex food chains require the high energetic efficiency of aerobic metabolism. And longer food chains were necessary for the evolution of larger organisms.

The requirement for a given minimum oxygen tension that can sustain aerobic metabolism is a function of size. The transport of O_2 into cells is limited by molecular diffusion and the external O_2 tension. The diffusive limitation increases with cell size. At the same time, the volume-specific energy metabolism tends to decrease with increasing cell size. With this background it is straightforward to show that for aerobic spherical cells the necessary external O_2 tension increases in proportion to $R^{1.25}$ where R is cell radius. In Figure 14.2 experimentally determined half-saturation constants for O_2 uptake (the O_2 tension that allows for 50% of the maximum respiratory rate) have been plotted against cell size for some unicellular eukaryotes (filled triangles) and for mitochondria (open triangle). Mitochondria can be considered representative of aerobic bacteria. (The O_2 requirement is expressed as half-saturation constants because these are rather well defined; maximum respiratory rates will be realized at about twice the half-saturation constant values.) It is seen that the

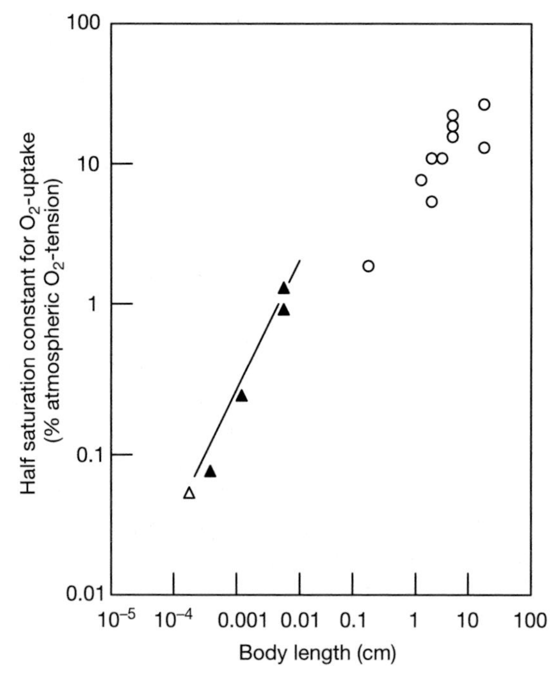

Fig. 14.2 The necessary external O_2 tension for maintenance of aerobic metabolism (expressed as half-saturation constant for O_2 uptake) as a function of size for different organisms. Open triangle: yeast mitochondria (representative for aerobic bacteria). Filled triangles: various unicellular protists. Open circles: various metazoa (an annelid worm and some crustaceans and fish). (Redrawn from Fenchel and Finlay, 1995.)

theoretical predictions are supported by the data. It is also seen that aerobic bacteria will cope with O_2 tensions that are well below 1% of the extant atmospheric level while larger protists require something around 1–2% of extant atmospheric O_2 saturation. This was the level attained about two billion years ago and it allowed for the diversification of aerobic unicellular protists.

In the case of multicellular animals, things are a bit more complex. Animals measuring more than a few millimetres solve the problem of oxygen supply to the cells with a circulatory system and blood with respiratory pigments (such as haemoglobin) so that oxygen can be distributed by advection—it is in part independent of molecular diffusion. The necessary external O_2 tension, however, still depends on body size. This is because the transport of oxygen to the mitochondria takes place through a number of steps, each of which depends on molecular diffusion. Thus oxygen must diffuse across the gill surfaces into the blood, from blood into tissue, from the tissue into cells, and eventually through the cytosol and into the mitochondria. Each of these steps requires differential concentrations in order to drive the diffusive fluxes, and more steps are required for larger organisms in order to expose the mitochondria to an O_2 tension of about 1% of atmospheric saturation. It is not, however, easy to make precise predictions about the relationship between body size and the required ambient O_2 tension. Animals may have different respiratory pigments with different oxygen affinities, different versions of the circulatory system, and other adaptations that complicate matters. However, Figure 14.2 shows that the general trend holds; the figure exemplifies this for an oligochaete worm and for various crustaceans and fish. It is seen that the largest animals require O_2 tensions that approach extant atmospheric saturation.

The Cambrian explosion—or 'Darwin's dilemma' (Chapter 2)—thus has a reasonable explanation. The evolution of aerobic respiration was the prerequisite for food chains, so unicellular protists could evolve and diversify during the late Precambrian. But the evolution of macroscopic animals required a substantially higher atmospheric oxygen tension (at least 10% of present atmospheric level). Once such values were reached there was scope for the development of larger organisms and the adaptive radiation that resulted in the zoological diversity we know today. Considered within the perspective of the four billion year period when there has been life on Earth, the origin of animals may seem like a rather late and exotic epi-phenomenon.

The events that led to the evolution of modern eukaryotic cells some two billion years ago and to multicellular organisms some 1.4 billion years later, do underscore contingencies of evolution. The origin of oxygenic photosynthesis and oxygen respiration was one condition, but without the special geochemical properties that allowed for the accumulation of oxygen in the atmosphere, we would all still be bacteria.

Further reading

The following is not a complete or comprehensive bibliography or reference list. It lists only some selected books and a few review papers. This should make it possible for readers to find the original literature on which this book is based.

Chapter 3 (and general)

The classic A. I. Oparin: *Origin of life* (English translation, The MacMillan Company, 1938; reprinted by Dover Publications, Inc., New York in 1953 and later) is still a readable book. It also provides a historical account of the controversy on spontaneous generation.

The corresponding paper by J. B. S. Haldane: *The origin of life*, was printed in *Rationalist Annual*, 1929, 148–69.

A general and excellent, if now somewhat outdated book is S. L. Miller and L. E. Orgel: *The origins of life on earth* (Prentice Hall, Inc., Englewood Cliffs, New Jersey, 1974). It includes a detailed treatment of experiments with prebiotic chemistry.

I. Fry: *The emergence of life on earth. A historical and scientific overview* (Rutgers University Press, New Brunswick, 2000) largely covers the first five chapters of the present book. The author is a philosopher and historian of science, but the treatment of biology is authoritative, the reference list is comprehensive and up to date, and it can be recommended.

C. Wills and J. Bada: *The spark of life. Darwin and the primeval soup* (Oxford University Press, Oxford, 2000) is a more popular approach to the subject. It includes much anecdotal stuff and emphasizes the more sensational aspects of the subject.

F. Dyson: *Origins of life* (2nd edn, Cambridge University Press, Cambridge, 1999) represents a physicist's approach. It is a slightly idiosyncratic defence of the 'metabolism first' approach to the origin of life, but it also contains many interesting ideas and viewpoints and can be recommended.

Chapter 4

E. Schrödinger's essay: *What is life* (1944; reprinted by Cambridge University Press, Cambridge, 1980) is still worth reading.

Two books by H. J. Morowitz: *Energy flow in biology* (Academic Press, New York, 1968) and *Beginnings of cellular life: metabolism recapitulates biogenesis* (Yale University Press, New Haven, 1992) should be consulted for a treatment of thermodynamics applied to biology.

Astrobiology is covered in P. W. Ward and D. Brownlee: *Rare earth: why complex life is uncommon in the universe* (Copernicus/Spinger–Verlag, New York, 2000). The reference list is comprehensive, but the book itself is somewhat wordy and contains a number of debatable statements.

M. Grady: *Astrobiology* (Smithsonian Institution Press, Washington DC) presents a very popular approach, but it includes pretty pictures from Mars, Yellowstone Park, and elsewhere.

A general, authoritative, and comprehensive textbook in (prokaryote) microbiology is *Brock biology of microorganisms*, 9th edn, by M. T. Madigan, J. M. Martinko, and J. Parker (Prentice-Hall, Inc., Upper Saddle River, New Jersey, 2000).

Chapter 5

The RNA world is thoroughly treated—theoretically as well as experimentally—in M. Eigen: *Steps towards life* (Oxford University Press, Oxford, 1992).

H. Gutfreund (ed.): *Biochemical evolution* (Cambridge University Press, Cambridge, 1981) includes an article on experimental RNA replication (and some other articles of relevance to Chapter 7 in the present book).

Chapter 6

The origin of genetic systems, the genetic code, and other problems pertaining to the origin of life is treated theoretically in J. Maynard Smith and E. Szathmáry: *The major transitions in evolution* (W. H. Freeman, Oxford, 1995). The book also discusses problems of the origin of eukaryotic cells (Chapter 8) and the origin of sex and of multicellular organisms (Chapters 9 and 10) and is highly recommended.

The origins of life: from birth of life to the origins of languages, by the same authors (Oxford University Press, Oxford, 1999) is a somewhat more popular and easily accessible book. Both books extend ideas about the origin of eukaryote cells, multicellularity, etc., to systems of even higher complexity such as the origin of social insects and human culture.

J. Monod: *Le hasard et la nécessité* (Édtions du Seuil, Paris, 1970) remains—in spite of its age—an interesting introduction to mechanisms of life and difficulties in understanding its origin.

Chapters 7, 11, and 14

Some of the books already mentioned also deal with microbial energy metabolism and its evolution (especially *Brock biology of microorganisms*).

The tropic is also treated in T. Fenchel and B. J. Finlay: *Ecology and evolution in anoxic worlds* (Oxford University Press, Oxford, 1995) and in T. Fenchel, G. M. King, and T. H. Blackburn: *Bacterial biogeochemistry* (Academic Press, San Diego, 1998).

J. F. Kasting: Earth's early atmosphere (*Science*, **29**, 920–6, 1993) is a recommend account of our understanding of the development of the Earth's atmosphere.

Chapters 8–10

L. Margulis: *Symbiosis in cell evolution* (2nd edn, W. H. Freeman and Company, New York, 1993) provides a vivid—in some parts a bit imaginative—treatment of the origin of eukaryotes as symbiotic consortia of prokaryote cells.

See also the above mentioned book by Maynard Smith and Szathmáry.

Problems of species concepts for bacteria have recently been discussed by F. M. Cohan: Bacterial species and speciation (*Systematic Biology*, **50**, 513–24, 2001).

Chapter 12

C. R. Woese: Bacterial evolution (*Microbiological Reviews*, **51**, 221–71, 1987) deals with the universal tree and microbial phylogeny based on rRNA gene sequences.

D. McL. Roberts, P. Sharp, G. Alderson, and M. Collins (ed.): *Evolution of microbial life* (Cambridge University Press, Cambridge, 1996) considers a number of problems related to the phylogeny of microbes, including more recent results based on gene sequencing.

Chapter 13

J. W. Schopf and C. Klein (ed.): *The proterozoic biosphere* (Cambridge University Press, Cambridge, 1992) represents a very comprehensive collection of articles on Precambrian life and the Precambrian biosphere (including descriptions of modern cyanobacterial mats).

An entertaining and readable description of Precambrian life and its geological exploration (but with a somewhat weak treatment of biology) is J. W. Schopf: *Cradle of life* (Princeton University Press, Princeton, 1999).

The development of biogeochemical cycling is provided in the above mentioned Fenchel, King, and Blackburn: *Bacterial biogeochemistry*.

The Gaia hypothesis is developed in J. E. Lovelock: *The ages of Gaia: a biography of our living earth* (2nd edn, W. W. Norton, New York, 1988).

Glossary

Adaptive radiation: The fact that some type of organism diverges into a number of lineages over a relatively short period of geological time.

Aerobic metabolism: Respiratory energy metabolism dependent on oxygen.

Alleles: Different versions of a gene with a given position in a chromosome.

Amoebae: Communal designation for a number of in part unrelated protist groups that are characterized by amoeboid motility. Amoebae occur in all kinds of aquatic habitats, in soils, and as parasites.

Anoxygenic photosynthesis: Photosynthetic processes occurring in different types of bacteria; oxygen is not produced as a metabolite.

Archaebacteria (archaea): A group of bacteria (prokaryotes) that in several important aspects deviate from the 'true' bacteria (eubacteria).

Assimilatory metabolism (= anabolic metabolism): The energy-requiring part of metabolism; it comprises the assimilation of molecules and the synthesis of biomolecules.

ATP, adenosine triphosphate: The all-dominating energy source in all kinds of cells for synthesis, transport across the cell membrane, motility, etc. The potential chemical energy is based on high energy phosphate bonds and the energy is released by splitting off one or two phosphate units.

Autocatalytic cycles: Cyclic chemical processes that result in the multiplication of one of the molecules entering the process for each cycle.

Biogeochemical cycles: Chemical processes on the Earth's surface (in the sea, on land, in the atmosphere) that are mediated by biological activity.

Biosphere: All living organisms and the chemical compounds that are transformed by biological activity.

Cell membrane: Membrane consisting of a double layer of lipid molecules in which different proteins are inserted. All cells are covered by a cell membrane. Almost all bacteria also possess a cell wall and an additional membrane; these are situated outside the (real) cell membrane.

Cell wall: Nearly all bacteria have a rigid cell wall outside of their cell membrane. Some types of eukaryotic cells also have cell walls.

Chlorophylls, bacteriochlorophylls: A type of biological molecule that are activated by light with an appropriate wavelength. Chlorophylls are green or greenish because they absorb red light; bacteriochlorophylls are almost colourless because they absorb near infrared light; they occur in bacteria with anoxygenic photosynthesis.

Ciliates: A large group of unicellular protists (protozoa), typically with hundreds of cilia on their surfaces. They occur in soil, in fresh- and seawater, and some live in or on other organisms. Ciliates measure between 10 μm and up to 1–2 mm.

Cryptomonads: A group of flagellates that include photosynthetic as well as heterotrophic forms.

Cyanobacteria: A group of eubacteria with oxygenic photosynthesis, formerly referred to as blue-green algae. Chloroplasts in green eukaryotes are descendants of endosymbiotic cyanobacteria. Cyanobacteria occur in all habitats exposed to light; macroscopically visible accumulations are common.

Cytochromes: Respiratory enzymes that play a role in the electron transport systems in connection with respiration and cyclic photophosphorylation.

Dinoflagellates: A large group of flagellates (eukaryotes) often equipped with a cellulose cell wall and two flagella. About half of them are photosynthetic; others feed on other protists, or in a few cases, they live as parasites.

Dissimilatory metabolism (= energy metabolism or catabolic metabolism): The part of metabolism that generates potential energy (ATP) through various chemo- or phototrophic processes.

Entropy: A concept that plays a role in the second law of thermodynamics (unit: joules/°K). In an isolated system, a process can occur only if it results in an increase in entropy. Another interpretation of entropy is as a measure of the probability of a given state of a system.

Eubacteria: Main group (domain) among the prokaryotes; includes most bacteria.

Eukaryotes: Organisms that are not bacteria (prokaryotes). Includes a large number of unicellular lineages and multicellular algae, plants, fungi, and animals.

Fermentation: Processes in which some organisms obtain energy by splitting large (organic) molecules into smaller ones.

Flagellates: An heterogeneous collection of unicellular eukaryote lineages; the representatives typically move using two (one to several) flagella.

Free energy (Gibbs free energy): A concept of thermodynamics (unit: Joule) that is particularly useful for describing the energetics of chemical processes. In order that a chemical process can proceed, the net change in free energy must be negative.

Genome: The genetic information (all genes) in an organism.

Genotype: The genetic information in an organism; the term is often used to specify a particular allele.

Gram-positive bacteria: Group belonging to the eubacteria and characterized by a special kind of cell wall. Includes among others the *Bacillus* species and the anaerobic, fermenting *Clostridia* species, both of which are common in soils. Both genera also include pathogenic forms. A single photosynthetic genus (*Heliobacter*) is also known.

Halophily: Requirement of a high salinity in the medium; extreme halophiles can grow in saturated brine.

Horizontal gene transfer: Direct transfer of genes (or genomes) from one organism to another.

Hydrolysis: Degradation of polymers into monomers (e.g. the degradation of proteins into amino acids) through uptake of H_2O molecules.

Isoprenoids: Branched hydrocarbons (built from isoprene units). Occur in the cell membrane of archaebacteria in the form of ether-bound glycerol esters.

Macroscopic/microscopic: Visible/non-visible with the naked eye. By 'micro-organisms' is usually understood unicellular eukaryotic or prokaryotic organisms even though a few may obtain macroscopic size.

Messenger RNA (mRNA): RNA molecules that are formed by transcription of DNA and are subsequently translated to protein in ribosomes.

Methanogenic bacteria: A type of archaebacteria that produce methane as metabolite.

Mitochondria: Cell organelle that occurs in almost all eukaryotic cells. The most essential parts of the energy metabolism take place in mitochondria. They are believed to descend from endosymbiotic proteobacteria.

Nucleoside: Molecule consisting of a carbohydrate (ribose in RNA, deoxyribose in DNA) and one of five organic bases. Nucleosides also occur as coenzymes together with different proteins and in other important biomolecules.

Nucleotide: Molecule consisting of a nucleoside and a phosphate molecule; nucleotides are the fundamental units of DNA and RNA.

Nucleotide sequence: The sequence of nucleotides in RNA or DNA molecules (genes).

Oxidation, oxidizing compounds: The loss of one or more electrons; oxidizing compounds tend to take up electrons, thus becoming reduced.

Oxygenic photosynthesis: A photosynthetic process in which H_2O is used as electron donor (for reducing CO_2 to organic matter) and with O_2 as metabolite. Occurs in cyanobacteria and in algae and plants.

Phagocytosis: Ability to take up particulate matter into cells; occurs only in eukaryotes. Phagotrophy means that organisms subsist on particulate matter for food.

Phenotype: All the properties in terms of structure and function of an organism that result from the genotype and from environmental influences.

Phospholipids: Lipid molecules combined with phosphate; they constitute the basal building block of the cell membranes of eubacteria and eukaryotes.

Photosynthesis, phototrophy: Processes by which organisms transform the energy of electromagnetic radiation into potential chemical energy.

Phylogeny: The relationship (in a genealogical sense) between different groups of organisms.

Plasmids: Small, circular pieces of extra-chromosomal DNA that may occur in bacteria. The distinction between virus and plasmids is not sharp.

Plate tectonics: The processes that move tectonic plates around on the Earth's surface and their geological consequences.

Polymers: Large molecules that consist of chains of identical or related smaller molecules (monomers). Examples include polypeptides and proteins that consist of amino acid units, nucleic acids that consist of nucleotides, and polysaccharides (like starch and cellulose) that are built from carbohydrates.

Prokaryotes: Bacteria including the eubacteria and the archaebacteria.

Proteobacteria: An important and diverse group within the eubacteria. Includes bacteria with anoxygenic photosynthesis (purple bacteria), bacteria with aerobic and anaerobic respiration (such as most sulphate reducers and denitrifying bacteria), and chemoautotrophic bacteria that oxidize reduced sulphur, iron, and nitrogen compounds. Among the aerobic heterotrophs there are several well-known pathogens.

Quasi-species: Designation for self-replicating RNA molecules that can be studied in the laboratory. Perhaps they can serve as a model for the origin of life.

Redox potentials: A solution of a mixture of a reduced and an oxidized version of a compound (e.g. the equilibrium $Fe^{3+} + e^- \leftrightarrow Fe^{2+}$) is characterized by an electrode potential (that in some cases can be measured directly with a platinum electrode against a reference potential), referred to as *Eh* with the unit of volts. A standard redox potential (E_o') is the electrode potential for a mixture of equal parts of the reduced and oxidized form at pH 7 and 20 °C relative to a hydrogen electrode at pH 0. It measures the tendency of the redox pair to oxidize or reduce other such redox pairs (see Figure 13.10). Standard redox potentials can be used to calculate free energy changes of redox processes.

Redox processes: Chemical processes that involves the transfer of electrons (oxidations, reductions).

Reduction, reducing chemical compounds: A molecule taking up one or more electrons; reducing compounds tend to deliver electrons to other compounds, thus being oxidized.

Respiration: A process in which organisms gain energy by oxidizing a substrate with an external electron acceptor (e.g. O_2).

Ribosomal RNA (rRNA): Two kinds of RNA + protein that together constitute a ribosome. Protein synthesis takes places in ribosomes in all kinds of organisms.

Sedimentary rocks: Rocks that have formed from material that has settled to the bottom of water bodies (lakes, the sea). Carbonate rock, shale, and sandstone are sedimentary rocks.

Spirochaetes: A group of filamentous, motile bacteria. Many free-living forms; a few are human pathogens (e.g. syphilis is caused by a spirochaete.)

Sulphate reducing (respiring) bacteria: Bacteria that make a living oxidizing low molecular weight organic compounds or hydrogen using sulphate as an electron acceptor, thus producing sulphide as metabolite. Widely distributed in the anaerobic sea bottom. Most belong to the proteobacteria, but they are also represented among other groups of eu- and archaebacteria.

Sulphur bacteria: Bacteria that oxidize reduced sulphur compounds (sulphide, elementary sulphur). Colourless sulphur bacteria are chemoautotrophs that use oxygen (or nitrate) as oxidant. Photosynthetic (green and purple) sulphur bacteria oxidize sulphur compounds in a photosynthetic process; these bacteria are typically anaerobes.

Syntrophy: Pairs of bacterial species with a complementary metabolism; the metabolite(s) of one species is the substrate(s) of the other species.

Thermophily: Requirement for high temperatures. Extreme thermophilic bacteria can grow at 85 °C under atmospheric pressure and up to 115 °C under hyperbaric conditions.

Transcription: The transfer of genetic information from DNA to mRNA molecules.

Transfer RNA (tRNA): Small RNA molecules that transport amino acids to the ribosomes.

Translation: Translation of the genetic information in mRNA to protein.

Index

Bold type denotes pages with illustrations.